Current Trends in Fisheries Biotechnology

Bijay Kumar Behera
Editor

Current Trends in Fisheries Biotechnology

Springer

Editor
Bijay Kumar Behera
College of Fisheries
Rani Lakshmi Bai Central Agricultural University
Jhansi, Uttar Pradesh, India

ISBN 978-981-97-3164-0 ISBN 978-981-97-3165-7 (eBook)
https://doi.org/10.1007/978-981-97-3165-7

This Springer imprint is published by the registered company Springer Nature Singapore Pte Ltd.
The registered company address is: 152 Beach Road, #21-01/04 Gateway East, Singapore 189721, Singapore

Foreword

It is heartening that the College of Fisheries, Rani Lakshmi Bai Central Agricultural University, Jhansi, Uttar Pradesh, India, is bringing out a book on *Current Trends in Fisheries Biotechnology*. It is remarkable that different branches of biotechnology have emerged from few disciplinary micro niches to well-defined dynamic research areas, making substantial impact in the field of fisheries science. The book covers emerging fields of genomics with a special focus on various biotechnological tools applied in fisheries science. Recent developments in metagenomics in aquatic ecosystem health surveillance, mitochondrial genes and their application in fish diversity studies, role of epigenetics, proteomics in fisheries and aquaculture, nanobiosensor technology applications for water quality monitoring, microplastics, bioremediation, eDNA, soil biotechnology and surrogacy technology in fisheries and aquaculture etc. have been included in the book. The contributors of this book are from reputed academic and scientific institutions across India. The editor has tried to capture such a wide and dynamic topic in a series of captivating articles highlighting newly emerging technologies in the field, protocols, advantages and potential future prospects with current trends in biotechnological research. I am confident that the publication shall be a reference book on the present status, trends and approaches of different biotechnological tools applied in fisheries and aquaculture. This publication will be beneficial to students and researchers in the field of biotechnology and related areas. I wish to compliment the editor and the contributors for their hard work and painstaking efforts.

Government of India, Department of Agricultural
Research and Education (DARE) and Indian
Council of Agricultural Research (ICAR),
Ministry of Agriculture and Farmers Welfare
New Delhi, India — Himanshu Pathak
27 February 2024

Preface

The book *Current Trends in Fisheries Biotechnology* mainly focuses on Mitochondrial Genes and their application in Fish Diversity studies. In recent times, most of the aquatic ecosystems are polluted due to anthropogenic activities. Therefore, metagenomics studies are important for aquatic ecosystem health surveillance. Similarly, studies on epigenetics in fisheries and aquaculture will be helpful for enhancing aquaculture production. Proteomics in fisheries and aquaculture, nanobiosensor technology applications for water quality monitoring, microplastics, bioremediation, eDNA, soil biotechnology and surrogacy technology in fisheries and aquaculture, etc., are the current area of research in fisheries and aquaculture biotechnology. These biotechnological applications can help to improve the productivity of farmed fish and other aquatic species and ensuring sustainable growth of the aquaculture industry. The book *Current Trends in Fisheries Biotechnology* contains 12 chapters and covers most of the topics related to fish biotechnology contributed by our distinguished scientists. The book's objective is to draw the attention of future budding scientists, researchers and policymakers for exploring biotechnological applications in fisheries and aquaculture for societal benefit.

Jhansi, Uttar Pradesh, India

Bijay Kumar Behera

Acknowledgements

The preparation of this book has been guided by several reputed scientists in Biotechnology and Fisheries Science. I am obliged to Dr. Himanshu Pathak, Secretary (DARE) and Director General (ICAR), New Delhi, Dr. Joykrushna Jena, Deputy Director General (Fisheries Science, ICAR), New Delhi, Dr. Ashok Kumar Singh, Vice Chancellor, Rani Lakshmi Bai Central Agricultural University, Jhansi, Padma Shri Dr. Subbanna Ayyappan, former Secretary (DARE) and Director General (ICAR), New Delhi, Dr. Trilochan Mohapatra, former Secretary (DARE) and Director General (ICAR), New Delhi, Prof. Shiva Dhar Singh, former Assistant Director General (Inland Fisheries, ICAR) and Dr. Basanta Kumar Das, Director, ICAR-Central Inland Fisheries Research Institute, Barrackpore, for providing their valuable inputs and support for this book. I thank the anonymous reviewers for their constructive comments that led to a substantial improvement in the quality of this book. I acknowledge all authors who have significantly contributed to this book. I also thank my wife Mrs. Jyotsna Dei and my son Shri Aman Jagannath for their constant motivation for bringing this book to reality. This work would not have been possible without support and enthusiasm from my colleagues, students and researchers, Dr. Ajaya Kumar Rout, Dr. Neelesh Kumar, Shri Partha Sarathi Tripathy, Shri Satya Narayan Parida and Shri Charan Singh Kushwaha. Finally, I also thank my publisher and its publishing editor, Springer Nature, for their continuous support and cooperation in the publication of this book.

I acknowledge Aakanksha Tyagi, Suraj Kumar, Muthuneela Muthukumar and Ms. Vijayalakshmi Dayalan from Springer Nature for their constant support and cooperation for the publication of this book.

Contents

Editor and Contributors

About the Editor

Bijay Kumar Behera is the Dean, College of Fisheries, Rani Lakshmi Bai Central Agricultural University, Jhansi, India. Dr. Behera has more than 21 years of research and teaching experience in the field of fish biotechnology, microbiology, bioinformatics and nanobiosensor technology. He has received Australian Government Endeavour Award, Japanese Government-Mombusho MEXT Fellowship, Prof. Har Swarup Award, Krushak Pratibha Award, Scientist of the Year Award, Dr. Hiralal Choudhury Award, Best Scientist Award and Dr. M. S. Swaminathan Best Indian Fisheries Scientist Award. He is also a Fellow of National Academy of Agricultural Sciences, Zoological Society of India and Inland Fisheries Society of India.

Contributors

Prachi Asgolkar Fish Genetics and Biotechnology, ICAR-Central Institute of Fisheries Education, Mumbai, Maharashtra, India

Rajib Bandyopadhyay Department of Instrumentation and Electronics Engineering, Jadavpur University, Kolkata, India

Tanushree Banerjee Aquatic Environmental Biotechnology and Nanotechnology Division, ICAR-Central Inland Fisheries Research Institute, Barrackpore, Kolkata, West Bengal, India

Bhaskar Behera Department of Biotechnology, Fakir Mohan University, Balasore, Odisha, India

Bijay Kumar Behera College of Fisheries, Rani Lakshmi Bai Central Agricultural University, Jhansi, Uttar Pradesh, India

Aquatic Environmental Biotechnology and Nanotechnology Division, ICAR-Central Inland Fisheries Research Institute, Barrackpore, Kolkata, West Bengal, India

Kampan Bisai College of Fisheries, Rani Lakshmi Bai Central Agricultural University, Jhansi, Uttar Pradesh, India

Jyotsna Dei Department of Instrumentation and Electronics Engineering, Jadavpur University, Kolkata, India

Manoharmayum Shaya Devi Aquatic Environmental Biotechnology and Nanotechnology Division, ICAR-Central Inland Fisheries Research Institute, Barrackpore, Kolkata, West Bengal, India

Shukla Devnath Department of Zoology, Vinoba Bhave University Hazaribagh, Hazaribagh, India

Aquatic Environmental Biotechnology and Nanotechnology Division, ICAR-Central Inland Fisheries Research Institute, Barrackpore, Kolkata, West Bengal, India

Diganta Dey Fish Genetics and Biotechnology, ICAR-Central Institute of Fisheries Education, Mumbai, Maharashtra, India

Sujata Dey Aquatic Environmental Biotechnology and Nanotechnology Division, ICAR-Central Inland Fisheries Research Institute, Barrackpore, Kolkata, West Bengal, India

Aquaculture Laboratory, Department of Zoology, The University of Burdwan, Burdwan, West Bengal, India

Souvik Dhar Aquatic Environmental Biotechnology and Nanotechnology Division, ICAR-Central Inland Fisheries Research Institute, Barrackpore, Kolkata, West Bengal, India

Koushik Ghosh Aquaculture Laboratory, Department of Zoology, The University of Burdwan, Burdwan, West Bengal, India

Pranab Gogoi Reservoir and Wetland Fisheries Division, ICAR-Central Inland Fisheries Research Institute, Barrackpore, West Bengal, India

Asim Kumar Jana Aquatic Environmental Biotechnology and Nanotechnology Division, ICAR-Central Inland Fisheries Research Institute, Barrackpore, Kolkata, West Bengal, India

Vikash Kumar Aquatic Environmental Biotechnology and Nanotechnology Division, ICAR-Central Inland Fisheries Research Institute, Barrackpore, Kolkata, West Bengal, India

Department of Zoology, Fakir Mohan University, Balasore, Odisha, India

Pritijyoti Majhi ICAR-Central Inland Fisheries Research Institute, Barrackpore, Kolkata, India

Aquatic Environmental Biotechnology and Nanotechnology Division, ICAR-Central Inland Fisheries Research Institute, Barrackpore, Kolkata, West Bengal, India

Chinmaya Kumar Parida Department of Biotechnology, Fakir Mohan University, Balasore, Odisha, India

Pranaya Kumar Parida ICAR-Central Inland Fisheries Research Institute, Barrackpore, Kolkata, India

Satya Narayan Parida Aquatic Environmental Biotechnology and Nanotechnology Division, ICAR-Central Inland Fisheries Research Institute, Barrackpore, Kolkata, West Bengal, India

Smruti Priyambada Pradhan Department of Community Medicine, IMS & SUM Hospital, Bhubaneswar, India

Sukanta Kumar Pradhan Department of Bioinformatics, Odisha University of Agriculture and Technology, Bhubaneswar, Odisha, India

Epari Venkata Rao Department of Community Medicine, IMS & SUM Hospital, Bhubaneswar, India

Kiran D. Rasal Fish Genetics and Biotechnology Division, ICAR-Central Institute of Freshwater Aquaculture, Bhubaneswar, Odisha, India

Fish Genetics and Biotechnology, ICAR-Central Institute of Fisheries Education, Mumbai, Maharashtra, India

Ajaya Kumar Rout Rani Lakshmi Bai Central Agricultural University, Jhansi, Uttar Pradesh, India

Aquatic Environmental Biotechnology and Nanotechnology Division, ICAR-Central Inland Fisheries Research Institute, Barrackpore, Kolkata, West Bengal, India

Suvra Roy Aquatic Environmental Biotechnology and Nanotechnology Division, ICAR-Central Inland Fisheries Research Institute, Barrackpore, Kolkata, West Bengal, India

Dhruba Jyoti Sarkar Aquatic Environmental Biotechnology and Nanotechnology Division, ICAR-Central Inland Fisheries Research Institute, Barrackpore, Kolkata, West Bengal, India

Siba Shinde Fish Genetics and Biotechnology, ICAR-Central Institute of Fisheries Education, Mumbai, Maharashtra, India

Jitendra Kumar Sundaray Fish Genetics and Biotechnology Division, ICAR-Central Institute of Freshwater Aquaculture, Bhubaneswar, Odisha, India

Partha Sarathi Swain Assistant Fisheries Officer, Office of Additional Fisheries Officer (Marine), Balasore, Odisha, India

Prajna Ritambhara Swain ICAR-Central Inland Fisheries Research Institute, Barrackpore, Kolkata, India

Prajnaritambhara Behera Swain Reservoir and Wetland Fisheries Division, ICAR-Central Inland Fisheries Research Institute, Barrackpore, West Bengal, India

Anjana Verma Department of Zoology, Vinoba Bhave University Hazaribagh, Hazaribagh, India

Metagenomics in Aquatic Ecosystem Health Surveillance

1

Bijay Kumar Behera and Ajaya Kumar Rout

Abstract

India has a lot of resources along its rivers that help millions of people get everything and services. But the health of these rivers is always in danger because of pollution and other human activities. The use of metagenomics-based methods in biological niches has made them very useful for keeping an eye on the health of rivers. Metagenomic studies are better than culture-based studies because they don't rely on errors that come from cultures. Next-generation sequencing (NGS) methods with high throughput and powerful data analysis software/tools could change the way metagenomics studies are done forever. High-throughput NGS platforms were used to look at sediment samples from India and a few river areas around the world to find out about the variety of microbes that live there and how they work. Access to more powerful computers has expanded the range of microbial biogeographic studies that look at how changes in the environment and human activities affect the community of bacteria. Researchers can now study and learn more about how the microbe community and the ecosystem around it connect better thanks to these cool and cutting-edge tools. They helped us figure out how the biogeochemical processes in river sediments work. This review piece lists the ways that metagenomics can be used to keep an eye on the health of rivers.

Keywords

Metagenomics · River ecosystem · Sediment sample · Microbial diversity

B. K. Behera (✉)
College of Fisheries, Rani Lakshmi Bai Central Agricultural University, Jhansi, Uttar Pradesh, India

A. K. Rout
Rani Lakshmi Bai Central Agricultural University, Jhansi, Uttar Pradesh, India

B. K. Behera (ed.), *Current Trends in Fisheries Biotechnology*,
https://doi.org/10.1007/978-981-97-3165-7_1

1.1 Introduction

The preservation of river health is crucial for the sustainable governance of inland aquatic ecosystems and fisheries. When assessing the health of a river, it is important to evaluate various aspects, including its physical, chemical, and biological characteristics. An ecologically healthy river should possess specific components that must remain within an acceptable range in order to sustain the operations of the riverine ecosystem. Generally, the evaluation of river water quality relies on a limited number of physical/chemical or biological characteristics. The Indian River system consists of 14 major rivers, 44 minor rivers, numerous tributaries, and branches, with a total length of 45,000 km. The riverine resources in this area sustain a significant amount of biodiversity and fisheries. It is believed that there are around 500 fish species, including commercially valuable ones, that thrive in these resources. They are being harvested for various purposes such as food, sports, ornamental purposes, as well as for natural seed and brooders for aquaculture hatcheries. Lately, nevertheless, the majority of rivers have experienced significant disturbances caused by human activities and are currently undergoing a crucial ecological transformation. The issues would be exacerbated and potentially more severe in the future as a result of the escalated rate of environmental deterioration caused by pollution and the upcoming climate change situation.

The microbiome is influenced by both environmental and human activities, and its ability to recover can be assessed by studying its taxonomic and functional makeup over time and space. Investigating microbial communities is essential for the health and long-term viability of the aquatic ecosystem, as well as for making informed decisions regarding its management and policies. Environmental genomics, also known as metagenomics, combined with next-generation sequencing (NGS) technology enables efficient and unbiased analysis of the makeup of aquatic microbial communities. Metagenomics is a technique that allows for the study of the microbiome without the need for culturing, by directly analyzing the genetic material retrieved from any given environment. The high-throughput reads obtained from metagenomics investigations are subsequently correlated with genes that are linked to microbial taxonomy and function, enabling the distinct capacity to characterize the microbial community found in nature. Assessing the variety and alteration of microorganisms is a valuable method for detecting, identifying, and predicting environmental conditions across different locations and time periods, as well as in reaction to human-induced influences. The metagenomic platform provides a significant chance to observe and analyze microbial communities and their essential functions. Due to these distinct advantages, it has gained popularity as a tool for evaluating aquatic microbial populations. As these environments undergo constant changes, it is believed that the impact on the microbial community can result in a decrease in microbial diversity, and alterations in microbial community function and health, ultimately leading to a decline in the quality and resilience of our ecosystems. This poses threats such as increased tolerance, persistence, antibiotic resistance, and the emergence of opportunistic pathogens. Metagenomics studies have been conducted to assess the extent of contamination caused by these hazardous substances in aquatic ecosystems, specifically in rivers and wetlands, which support the livelihood of numerous aquatic organisms and plants. Shotgun metagenomics is the indiscriminate sequencing of the genetic material present in a microbiome. This technique has greatly transformed the field of research by leveraging improvements in sequencing technology, which allow for higher throughput and the capacity to identify both genes and the composition of the microbiome.

The microbiological composition of river water is diverse and includes bacteria, viruses, microeukaryotes, fungi, and archaea. These microorganisms accumulate in rivers from the surrounding upstream area through various means such as overland flow, groundwater flow, aerial deposition, and streamflow. They can be either attached to particles or exist independently (Wetzel 2001). Upon entering the water column, bacteria can either continue downstream or adhere to the river bed or banks and form biofilms. Biofilms can serve as a reservoir of microorganisms in the water column through the shedding of cells. Numerous species undergo a transition from being suspended in water ("planktonic") to being linked with the river bed ("benthic"), and vice versa, at different stages of their life cycle or due to changes in the surrounding environment (Sigee 2005). The structure of microbial communities in river water is influenced by environmental factors, which in turn affect the species residing in the river, the organisms residing in the catchment area, and the mechanisms of transportation.

Pristine rivers are essential for the well-being of both humans and the environment. Rivers serve as a significant reservoir of potable water and play a crucial role in facilitating agricultural irrigation, supporting fisheries, and offering recreational opportunities. Human activities can undermine these functions by reducing the overall well-being of rivers through several means, including introducing harmful substances like fertilizers, metals, pesticides, and industrial solvents, as well as introducing harmful microorganisms like fecal coli, diseases, and invasive species (Meybeck 2003). Microbial community alterations can serve as a direct indicator of microbial pollution, as well as a means to monitor chemical pollution. For instance, the presence of algal bloom could suggest chemical contamination resulting from the runoff of fertilizers into a river, whereas an upsurge in bacteria tolerant to metals could indicate contamination by metals. Alterations in river microbial populations can serve as an indication of contamination, either from direct water pollution or from the catchment region if it impacts the microorganisms entering the river. Microbial communities in rivers not only detect the existence of contaminants but also have the ability to modify them. When the levels of nitrate or nitrite are elevated, bacteria have the ability to utilize these molecules as electron acceptors in place of oxygen. This process, known as denitrification, leads to the permanent elimination of reactive nitrogen (Findlay 2010). These processes can influence the effect of nutrient contamination on the downstream receiving waterways.

River microbial communities play a crucial role in global nutrient cycles by converting inorganic nutrients and releasing nutrients through the decomposition of organic materials (Meybeck 2003; Findlay 2010). Microorganisms can elevate phosphate concentrations in the water column by decomposing organic sediments and directly releasing microbial metabolites (Sigee 2005). Microorganisms also absorb inorganic phosphorus and nitrogen directly from the water column (as discussed in Findlay 2010). These mechanisms facilitate the availability of nutrients to other species and promote the proliferation of microorganisms, hence helping the consumers further up in the food chain.

1.2 Advantages of Metagenomic Studies of Environmental Samples

Historically, culture-based approaches have been the norm when investigating microbiomes from various ecological niches. It is believed that culturable bacteria comprise less than 1% of the total bacterial species in environmental communities (Amann et al. 1995). In order to address the constraints associated with conventional culture-based methodologies, Handelsman et al. (1998) proposed a metagenomic approach as a viable and auspicious alternative for characterizing the DNA of the entire microbial community. The utilization of this methodology to analyze millions of sequence reads in order to classify thousands of species has unveiled an abundance of taxonomic and functional variety in both terrestrial and aquatic environments (Sogin et al. 2006). According to Staley et al. (2014), these techniques offer a plausible substitute for conventional culture-based or molecular approaches when evaluating the well-being of ecosystems and public health hazards. The metagenomic investigations conducted by various researchers are detailed in Table 1.1.

Table 1.1 Metagenomic studies carried out in different rivers

River	Sample type	Identification of microbes/genes	References
Amazon	Water	Classified the 16S rRNA gene fragments recovered from the metagenome (722 sequences)	Ghai et al. (2011)
Yamuna	Water	Identified metabolic and antibiotic resistance genes	Mittal et al. (2019)
Bogotá	Water	Analyzed metagenome level metabolic network	Ruiz-Moreno et al. (2019)
Ganga	Water and sediment	Identified 175 and 247 bacterial genera involved to hosts ARGs MRGs respectively in water and sediment	Reddy and Dubey (2019)
James	Water and sediment	Identified planktonic microbial consortia	Brown et al. (2015)
Yamuna	Sediment	Identified 139 antimicrobial resistance (AMR) genes from 39 microbial species	Das et al. (2020)
Yamuna	Sediment	Identified antibiotic resistance genes	Mittal et al. (2019)
Ganga and Yamuna	Sediment	Identified 69 helpful bacteria	Behera et al. (2020a, b)
Ganga and Yamuna	Sediment	Identified bioremediating bacteria of 45 genera with 92 species and fungi belonging to 13 genera with 24 species	Behera et al. (2020a, b)

1.3 Bioinformatics Analysis of Metagenomic Data

The advent of next-generation sequencing (NGS) technologies has brought about a revolutionary approach developing computational techniques for analyzing the vast amounts of data obtained from community-based profiling of microorganisms (Fig. 1.1). QIIME, also known as Quantitative Insights into Microbial Ecology, is a widely used software application for the processing and analysis of taxonomic marker genes (Caporaso et al. 2010). The Metagenomic Analyzer (MEGAN) is a tool used for analyzing shotgun sequences. It overcomes the challenges of assembling vast sequence data from environmental samples and the scarcity of phylogenetically important marker genes (Huson et al. 2007). The Community Cyberinfrastructure for Advanced Marine Microbial Research and Analysis (CAMERA) is an online library that combines established and developing bioinformatics tools to analyze metagenomic data. It includes analytical packages and workflows (Seshadri et al. 2007). The program conducts a comparative analysis of sequence reads with a reference database and generates the results for further examination through a graphical user interface. Web servers such as MG-RAST provide the ability to upload and retain data, with automatic taxonomic and functional annotations (Meyer et al. 2008). These programs have integrated bioinformatics techniques to enable comprehensive and intricate investigation of ecological datasets.

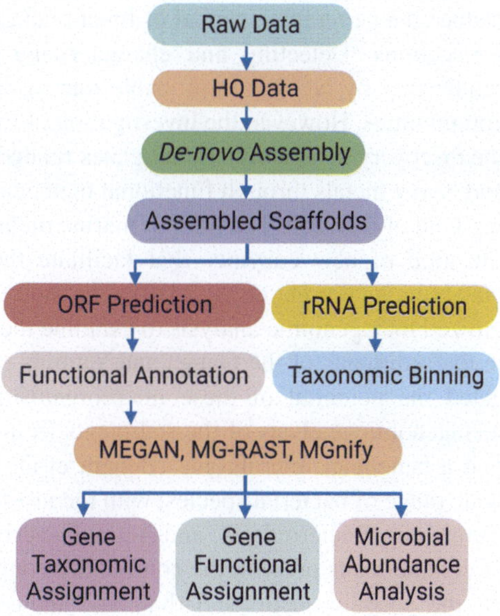

Fig. 1.1 Bioinformatics pipeline for metagenomic data analysis

1.4 Metagenomic Characterization of Functional Diversity

Metagenomics is transforming the field of aquatic resource management by offering extensive knowledge about microbial communities and their roles. Advancements in sequencing technologies and bioinformatics are improving our capacity to monitor, comprehend, and control aquatic environments. To effectively tackle environmental concerns and promote sustainable use of aquatic resources, it is crucial to prioritize standardizing methodology, performing long-term studies, and integrating metagenomic data into management and policy frameworks in future endeavors (Behera et al. 2021a). Metagenomics is becoming more commonly employed to do thorough evaluations of water quality. This technology detects microbial pollutants, antibiotic-resistant genes, and possible pathogens, providing a comprehensive assessment of water quality. Scientists employ metagenomics to investigate the influence of contaminants, such as heavy metals, hydrocarbons, and microplastics, on microbial populations. This aids in comprehending the ecological ramifications of pollution. Metagenomic investigations offer valuable insights into the variety and functional capacity of microbial communities, serving as markers of their overall health and ability to recover ecosystems. Identifying certain microbial strains and enzymes with the ability to break down contaminants is crucial for the development of bioremediation techniques aimed at purifying polluted water bodies. Research efforts are directed toward examining the impact of climate change on microbial communities, specifically their response to alterations in environmental variables such as temperature and salinity shifts (Behera et al. 2021a). In focusing on the essential sets of genes that are shared by prokaryotic lineages, we can develop a phylogenetic link based on genes that have functional significance. This approach is similar to constructing phylogenetic trees using sequences from related taxa (Segata and Huttenhower 2011). This indicates that taxonomy data can also be utilized to deduce the geographic spread of functional genes by examining their phylogenetic connections. Detecting and characterizing new genes utilizing whole-genome sequencing by NGS is improbable due to the intricate composition of microbial communities. However, the investigation of functional trait distribution patterns and the discovery of new functional genes related to traits such as antibiotic resistance and heavy metals through functional metagenomic studies continue to be a promising field of research. The field of marine metagenomics has contributed to the identification of new enzymes that facilitate the production of potentially valuable metabolites (Barone et al. 2014). The study conducted by Behera et al. (2020a) utilized metagenomic analysis to examine the bacterial and fungal diversity present in the sediments of the Ganga and Yamuna rivers in India. The study also investigated the potential of these microorganisms for bioremediation purposes. The metagenomic analysis of the sediments from the rivers Ganga and Yamuna identified a beneficial microbiome (Behera et al. 2020b). The river Yamuna harbors a wide range of bacterial species, with the most prevalent phyla being Proteobacteria, Firmicutes, Bacteroidetes, and Actinobacteria. According to the quantity and type of contaminants present, the community composition of the river varies depending on the specific sections. Urban and industrial regions generally exhibit more microbial diversity as a result of the wide range of pollutants involved. The contaminated areas frequently contain potentially harmful bacteria such as Escherichia coli, Vibrio spp., and Aeromonas spp., which can cause health hazards for both humans and animals. The metagenomic study shows that the microbes have a lot of different metabolic pathways. This suggests that they might be able to break down different kinds of organic pollutants, like pesticides, heavy metals, and hydrocarbons. The fact that many genes are involved in breaking down xenobiotics, like polycyclic aromatic hydrocarbons and chlorinated chemicals, shows that the microbes are very important for cleaning up polluted environments. Although there is pollution, the fact that there are functional genes linked to cycling nitrogen, sulfur, and phosphorus shows that biogeochemical processes are still going on. There is a notable abundance of antibiotic resistance genes (ARGs) in the contaminated areas. These genes provide resistance to beta-lactams, tetracyclines, sulfonamides, and other antibiotics. Efflux pumps, antibiotic-degrading enzymes, and target alteration are some of the resistance mechanisms. This means that there is a high chance that bacteria will become resistant. The elevated abundance of antibiotic resistance genes (ARGs) gives rise to apprehensions regarding the dissemination of antibiotic resistance, which has the potential to impact both environmental and public health. In a recent metagenomic investigation conducted by Das et al. (2020) has successfully found antibiotic resistance genes in the sediments of the river Yamuna. The sediments of the river Ganges contain a wide range of microorganisms, such as bacteria, archaea, fungi, and viruses, which contribute to its high microbial diversity. An analysis of the metagenomic composition of sediments in the river Ganges has provided valuable information on the variety of microorganisms present and the existence of enzymes that can break down plastics and other foreign substances. Proteobacteria, Bacteroidetes, Actinobacteria, and Firmicutes are among the most prevalent bacterial phyla. These phyla are recognized for their wide range of metabolic capabilities and their capacity to adjust to diverse environmental situations (Rout et al. 2022; Choudhury et al. 2023). The metagenomic data reveals a diverse range of functional genes associated with different metabolic pathways, such as nitrogen fixation, sulfur metabolism, and organic matter degradation. Xenobiotic-degrading Enzymes such as monooxygenases, dioxygenases, and hydrolases that participate in the degradation of several xenobiotics (artificial substances not naturally occurring in the environment) were detected. It was observed that there are complete degradation pathways for pollutants such as polycyclic aromatic hydrocarbons (PAHs), insecticides, and other harmful substances. The sediments had a high abundance of microorganisms that have the ability to break down xenobiotics, including those belonging to the Sphingomonas, Burkholderia, and Mycobacterium genera. Genes associated with the response to oxidative stress, tolerance to heavy metals, and other environmental stressors are widespread, suggesting that microorganisms have adapted to contaminated circumstances (Behera et al. 2021b; Rout et al. 2024a). An in-depth examination of the sediment samples from the river Ganges in India by metagenomic analysis offers a thorough understanding of the presence of antibiotic-resistant genes and virulence factors. The identification of antibiotic resistance genes (ARGs) and virulence factors in the sediments suggests possible threats to public health, especially for communities that depend on the river for drinking water, irrigation, and recreational purposes. There is a notable abundance of antibiotic-resistant genes, including those that provide resistance to frequently used antibiotics including beta-lactams, tetracyclines, aminoglycosides, and sulfonamides. The resistance mechanisms that have been identified include efflux pumps, enzymes that inactivate antibiotics (such as beta-lactamases), and alteration of antibiotic targets. The widespread occurrence of antibiotic resistance genes (ARGs) presents a significant danger for the transmission of antibiotic resistance via horizontal gene transfer, which could have adverse effects on both human and animal well-being (Rout et al. 2023; Rout et al. 2024a). The sediments include genes that encode virulence agents, including poisons, adhesins, and invasion proteins. The potentially harmful bacteria that have been identified belong to the genera Escherichia, Vibrio, and Aeromonas. These bacteria are known to cause diseases in both people and animals. The existence of virulence factors implies that the sediments in the Ganges could act as reservoirs for pathogens, therefore contributing to the occurrence of waterborne diseases (Rout et al. 2023; Rout et al. 2024b). An examination of the variety of bacteriophages in the river Ganges offers valuable knowledge about their possible function in controlling harmful bacteria in this significant water-based ecosystem. The sediments and water of the river Ganges harbor a wide array of bacteriophages (phages), which are viruses that invade bacteria. Phages belonging to several families, such as Myoviridae, Siphoviridae, and Podoviridae, have been discovered. These families are renowned for their extensive range of hosts and diverse infection methods. Phages display selectivity towards different bacterial hosts, encompassing both environmental bacteria and possible human diseases. Phages that target dangerous bacteria, including Escherichia coli, Vibrio cholerae, and Salmonella spp., have been discovered in the river Ganges. These bacteriophages have the ability to decrease the populations of these pathogenic bacteria. The found phages have the potential to be utilized in phage treatment applications, providing a natural and focused method to manage bacterial infections in both human and veterinary settings. Phages act as innate biocontrol agents, regulating bacterial populations and promoting microbial equilibrium in aquatic ecosystems. Phages contribute to biogeochemical cycles, such as carbon and nitrogen cycling, by releasing nutrients into the environment through the lysis of bacterial cells. Metagenomic investigations unveil the genetic variability of phages, encompassing genes implicated in host identification, replication, and cell lysis. The presence of diverse phages is crucial for comprehending phage ecology and evolution. Phages play a role in promoting the transfer of genes between bacteria, which in turn affects the dynamics and development of microbial communities. This encompasses the transmission of advantageous genes, such as those that provide resistance to antibiotics or enhance metabolic capacity. The existence of bacteriophages that specifically attack disease-causing bacteria in the river Ganges indicates the possibility of reducing waterborne illnesses. Bacteriophages have the potential to be utilized as therapeutic interventions or prophylactic strategies to combat bacterial epidemics. Phages have the potential to be employed in bioremediation techniques aimed at decreasing bacterial contamination in the river, thereby enhancing water quality and promoting the health of the ecosystem. Additional investigation into the interactions between phages and bacteria, as well as the therapeutic capabilities of phages found in the Ganges, can aid in the development of alternate treatments to antibiotics. This is especially important given the increasing prevalence of antibiotic resistance (Behera et al. 2023).

1.5 Conclusion

This review emphasizes the effectiveness of utilizing metagenomics tools to analyze environmental samples in order to accurately assess diversity, biogeography, potential public health risks, and anthropogenic influences. The creation of novel computer methods for analyzing data from metagenomics investigations has revealed the identification of previously unknown microbes, some of which may possess industrial uses. A comprehensive identification and characterization of these microbial communities will provide more accurate conclusions of the behaviors that have a substantial impact on shaping the structure of bacterial communities. Furthermore, the utilization of high-throughput technology alongside advanced computational resources in metagenomic, metatranscriptomic, metabolomic, and metaproteomic research will enable microbial ecologists to address these gaps in knowledge. Consequently, this will provide a deeper understanding of the relationship between the environment, microbial communities, and human activities.

References

Amann RI, Ludwig W, Schleifer KH (1995) Phylogenetic identification and in situ detection of individual microbial cells without cultivation. Microbiol Rev 59(1):143–169

Barone R, De Santi C, Palma Esposito F, Tedesco P, Galati F, Visone M, Di Scala A, De Pascale D (2014) Marine metagenomics, a valuable tool for enzymes and bioactive compounds discovery. Front Mar Sci 1:38

Behera BK, Chakraborty HJ, Patra B, Rout AK, Dehury B, Das BK, Sarkar DJ, Parida PK, Raman RK, Rao AR, Rai A (2020a) Metagenomic analysis reveals bacterial and fungal diversity and their bioremediation potential from sediments of river Ganga and Yamuna in India. Front Microbiol 11:2531

Behera BK, Patra B, Chakraborty HJ, Sahu P, Rout AK, Sarkar DJ, Parida PK, Raman RK, Rao AR, Rai A, Das BK (2020b) Metagenome analysis from the sediment of river Ganga and Yamuna: in search of beneficial microbiome. PLoS One 15(10):e0239594

Behera BK, Dehury B, Rout AK, Patra B, Mantri N, Chakraborty HJ, Sarkar DJ, Kaushik NK, Bansal V, Singh I, Das BK (2021a) Metagenomics study in aquatic resource management: recent trends, applied methodologies and future needs. Gene Reports 25:101372

Behera BK, Sahu P, Rout AK, Parida PK, Sarkar DJ, Kaushik NK, Rao AR, Rai A, Das BK, Mohapatra T (2021b) Exploring microbiome from sediments of River Ganga using a metagenomic approach. Aquat Ecosyst Health Manag 24(4):12–22

Behera BK, Patra B, Chakraborty HJ, Rout AK, Dixit S, Rai A, Das BK, Mohapatra T (2023) Bacteriophages diversity in India's major river Ganga: a repository to regulate pathogenic bacteria in the aquatic environment. Environ Sci Pollut Res 30(12):34101–34114

Brown BL, LePrell RV, Franklin RB, Rivera MC, Cabral FM, Eaves HL, Gardiakos V, Keegan KP, King TL (2015) Metagenomic analysis of planktonic microbial consortia from a non-tidal urban-impacted segment of James River. Stand Genomic Sci 10:65

Caporaso JG, Kuczynski J, Stombaugh J, Bittinger K, Bushman FD, Costello EK, Fierer N, Pena AG, Goodrich JK, Gordon JI, Huttley GA (2010) QIIME allows analysis of high-throughput community sequencing data. Nat Methods 7(5):335–336

Choudhury N, Sahu TK, Rao AR, Rout AK, Behera BK (2023) An improved machine learning-based approach to assess the microbial diversity in major North Indian River ecosystems. Gene 14(5):1082

Das BK, Behera BK, Chakraborty HJ, Paria P, Gangopadhyay A, Rout AK, Nayak KK, Parida PK, Rai A (2020) Metagenomic study focusing on antibiotic resistance genes from the sediments of river Yamuna. Gene 758:144951

Findlay S (2010) Stream microbial ecology. J N Am Benthol Soc 29(1):170–181

Ghai R, Rodàíguez-Valera F, McMahon KD, Toyama D, Rinke R, de Oliveira TCS, Garcia JW, de Miranda FP, Henrique-Silva F (2011) Metagenomics of the water column in the pristine upper course of the Amazon river. PLoS One 6(8):e23785

Handelsman J, Rondon MR, Brady SF, Clardy J, Goodman RM (1998) Molecular biological access to the chemistry of unknown soil microbes: a new frontier for natural products. Chem Biol 5(10):R245–R249

Huson DH, Auch AF, Qi J, Schuster SC (2007) MEGAN analysis of metagenomic data. Genome Res 17(3):377–386

Meybeck M (2003) Global analysis of river systems: from Earth system controls to Anthropocene syndromes. Philos Trans R Soc Lond B Biol Sci Ser 358(1440):1935–1955

Meyer F, Paarmann D, D'Souza M, Olson R, Glass EM, Kubal M, Paczian T, Rodriguez A, Stevens R, Wilke A, Wilkening J (2008) The metagenomics server—a public resource for the automatic phylogenetic and functional analysis of metagenomes. BMC Bioinformatics 9:386

Mittal P, Prasoodanan PKV, Dhakan DB, Kumar S, Sharma VK (2019) Metagenome of a polluted river reveals a reservoir of metabolic and antibiotic resistance genes. Environ Microbiome 14(1):5

Parida PK, Behera BK, Dehury B, Rout AK, Sarkar DJ, Rai A, Das BK, Mohapatra T (2022) Community structure and function of microbiomes in polluted stretches of river Yamuna in New Delhi, India, using shotgun metagenomics. Environ Sci Pollut Res 29(47):71311–71325

Reddy B, Dubey SK (2019) River Ganges water as reservoir of microbes with antibiotic and metal ion resistance genes: high throughput metagenomic approach. Environ Pollut 246:443–451

Rout AK, Dehury B, Parida PK, Sarkar DJ, Behera B, Das BK, Rai A, Behera BK (2022) Taxonomic profiling and functional gene annotation of microbial communities in sediment of river Ganga at Kanpur, India: insights from whole-genome metagenomics study. Environ Sci Pollut Res 29(54):82309–82323

Rout AK, Tripathy PS, Dixit S, Behera DU, Behera B, Das BK, Behera BK (2023) Unveiling the microbiome landscape: a metagenomic study of bacterial diversity, antibiotic resistance, and virulence factors in the sediments of the river Ganga. India Antibiotics 12(12):1735

Rout AK, Dixit S, Tripathy PS, Rout SS, Parida SN, Parida PK, Sarkar DJ, Das BK, Singh AK, Behera BK (2024a) Metagenomic landscape of sediments of river Ganga reveals microbial diversity, potential plastic and xenobiotic degradation enzymes. J Hazard Mater 471:134377

Rout AK, Tripathy PS, Dixit S, Behera DU, Behera B, Das BK, Behera BK (2024b) Metagenomics analysis of sediments of river Ganga, India for bacterial diversity, functional genomics, antibiotic resistant genes and virulence factors. Curr Res Biotechnol:100187

Ruiz-Moreno HA, López-Tamayo AM, Caro-Quintero A, Husserl J, Barrios AFG (2019) Metagenome level metabolic network reconstruction analysis reveals the microbiome in the Bogotá River is functionally close to the microbiome in produced water. Ecol Model 399:1–12

Segata N, Huttenhower C (2011) Toward an efficient method of identifying core genes for evolutionary and functional microbial phylogenies. PLoS One 6(9):e24704

Seshadri R, Kravitz SA, Smarr L, Gilna P, Frazier M (2007) CAMERA: a community resource for metagenomics. PLoS Biol 5(3):e75

Sigee D (2005) Freshwater microbiology: biodiversity and dynamic interactions of microorganisms in the aquatic environment. Wiley, New York

Sogin ML, Morrison HG, Huber JA, Welch DM, Huse SM, Neal PR, Arrieta JM, Herndl GJ (2006) Microbial diversity in the deep sea and the underexplored "rare biosphere". Proc Natl Acad Sci 103(32):12115–12120

Staley C, Gould TJ, Wang P, Phillips J, Cotner JB, Sadowsky MJ (2014) Bacterial community structure is indicative of chemical inputs in the Upper Mississippi River. Front Microbiol 5:524

Wetzel RG (2001) Limnology: lake and river ecosystems. Gulf Professional Publishing, Hoston

Mitochondrial Genes and Their Application in Fish Diversity Studies

2

Satya Narayan Parida, Chinmaya Kumar Parida, Ajaya Kumar Rout, Vikash Kumar, Souvik Dhar, Kampan Bisai, Bhaskar Behera, and Bijay Kumar Behera

Abstract

Developing an evaluation for the identification and preservation of valuable species is crucial for addressing the management and conservation of biodiversity. The use of the mitochondrial genome as a molecular marker has the potential to identify the species. Statistical analysis of molecular markers transformed the analytical capability required to investigate genetic diversity. The study on mitochondrial genes like COI, cytochrome-b, and D-loop delivers various scientific insights like an assessment between wild and hatchery populations, population structure, species identification, and genetic variation which have importance in the practice of aquaculture practice. In this chapter, we provide some insights related to the mitochondrial genes and their role in structural variation in fish diversity.

Keywords

Mitochondrial genes · Biodiversity · D-loop · Genetic variation · Aquaculture

S. N. Parida · A. K. Rout · V. Kumar · S. Dhar
Aquatic Environmental Biotechnology and Nanotechnology Division, ICAR-Central Inland Fisheries Research Institute, Barrackpore, Kolkata, West Bengal, India

C. K. Parida · B. Behera
Department of Zoology, Fakir Mohan University, Balasore, Odisha, India

K. Bisai · B. K. Behera (✉)
College of Fisheries, Rani Lakshmi Bai Central Agricultural University, Jhansi, Uttar Pradesh, India

B. K. Behera (ed.), *Current Trends in Fisheries Biotechnology*,
https://doi.org/10.1007/978-981-97-3165-7_2

2.1 Introduction

Fishes are regarded as the world's cheapest sources of animal protein and a wide variety of minerals. One of the richest nations in terms of fish genetic resources, which are dispersed over the network of rivers, is India, which is habitat to about 11% of all known species of fish in the world (Kali Mahapatra and Kumar Sarkar 2015). The development of accurate and dependable tools for the quick screening of DNA sequence variation is one of the major accomplishments of contemporary biology (Yang et al. 2014). Newly, several molecular techniques have been applied to study the genetic variability among species, these are AFLP (amplified fragment length polymorphism), RAPD (random amplified polymorphic DNA), RFLP (restriction fragment length polymorphism), and using nuclear mitochondrial DNA (Yang et al. 2013). The mitochondria are an organelle in the cell that may directly turn organic materials into energy to maintain the biological activity of cells as it is the primary site for ATP synthesis through the Krebs cycle and oxidative phosphorylation in eukaryotes (Sun et al. 2021). mtDNA is self-replicative and semi-conservative with a closed circular double-stranded structure (Castro Paz et al. 2014). In eukaryotes, mtDNA is regarded as the secondary genetic information system of eukaryotic cells (Ruan et al. 2020). Fish mtDNA sequencing has been made easier by recent developments in molecular biology, such as second-generation sequencing technology. This has improved our understanding of the patterns of fish mitogenomes (Alam et al. 2014). Mitogenomes have become useful tools in evolutionary and phylogenetic studies of fishes as mtDNA sequences of a number of fish species have been identified successfully (Wu et al. 2019). Mitochondrial DNA provides new information to study the wild and culturable vertebrate fish species. The utilization of mitochondria will come to mind because the inheritance pattern of mitochondrial DNA occurs without genetic recombination. mtDNA is used as a biomarker system in phylogenetic and population studies (Ladoukakis and Zouros 2017). The genetic resources are mainly important as they help to identify traits and genes thus providing the basis on which to differentiate between species. Characterization is defined as the quality or character of an organism, and it also describes the character or the uniqueness of an organism which makes it different from the other organism and it follows the Mendelian principle that involves a specific DNA sequence. The mitochondria are found in the cells of every eukaryotic organism. Mammalian cells contain around 1–2% of their total DNA in the form of double-stranded mitochondrial DNA, which is around 16 kb in length. The mitochondrial genome codes for 13 important oxidative phosphorylation subunit polypeptides/proteins are the following: 7 Complex I subunits (ND4L and ND1–6), 1 Complex III subunit (Cytb), 2 Complex V subunits (ATPase 6 and 8), and 3 Complex IV subunits (COI-III) along with that 22 tRNAs and 2 rRNAs (12S and 16S rRNA) which are also encoded both of which are necessary for mitochondrial protein synthesis (Attardi 1988; Pirger et al. 2009; Wallace 1999). The DNA replication, transcription, and translation machinery in mitochondria is organelle-specific (Anderson et al. 1981).

Developing an evaluation to identify and preserve valuable species is crucial for addressing the management and conservation of biodiversity (Faith 1994; Liu et al. 1999). Before the characterization was made by analyzing mitochondrial DNA other conventional methods like an immunological procedure was used as immunological precipitation of antigen fraction using a corresponding antibody raised against the soluble protein. But this method bears so many drawbacks as the stability of antibodies as well as insufficient commercial support (Preziosi 2007). It is an unspecific technique as it represents the family that is producing the same antibody. The difference within a species level shows the adaptation of a species toward the changing environment as well as other evolutionary mechanisms, such as natural selection and genetic drift, emerges between various species, resulting in species divergence and ensuring the survival of the species (Gregory 2009). In the research field, genetic diversity data is usually used to study evolution, genetic improvement programs, and conservation and management of natural resources. Here the molecular marker plays a role in facilitating genetic studies of desired species. The molecular marker data combined with the new statistical programs enhances the critical power to explore genetic diversity. There are various molecular methods based on DNA and protein. Based on DNA, both the mitochondrial (*mt*DNA) as well as nuclear DNA is used (Shirasawa and Hirakawa 2013). The methods like microsatellites, AFLP (amplified fragment length polymorphism), RAPD (random amplification of polymorphic DNA), RFLP (restriction fragment length polymorphism), and SNP (single nucleotide polymorphism) are used mostly in the field of aquaculture like fisheries. These biomarkers provide numerous observations which have importance in aquaculture such as (1) species recognition, (2) population structure study and genetic characterization in farm-raised and wild populations, (3) alternation among wild type and cultivable populations, (4) assessment of demographic bottleneck in natural population, (5) propagation directed reintegration programs. There are two types of markers, i.e., type-I and type-II markers. Type-I markers are related to the gene having a known function whereas type-II markers are related to any region in the DNA. Allozyme belongs to the type-I marker because the encoded protein has a previously known function whereas RAPD and microsatellite markers belong to the type-II marker due to unknown function (Tanya and Kumar 2010). Among both RAPD and microsatellite markers, the RAPD marker amplifies any unknown genomic region by PCR. Though microsatellite belongs to type-II marker they are associated with the gene having a known function. Type-I markers generally serve as a connection to the transfer of character from a rich-fit species to poor-fit species. Type-II markers such as RAPD, AFLP, and microsatellite are broadly used in studies of population genetics to map the diversity among the fish population. In recent times the use of mtDNA analysis has increased day by day to study the phylogeny and population of an organism as it is maternally inherited without having any genetic recombination. In vertebrate species, the sequence divergence is more rapid in mtDNA than that in nuclear DNA. This divergence is accredited to a rapid mutation rate without having a repair mechanism at the time of replication (Merheb et al. 2019). Since there are several copies of mtDNA present in each cell, it is simpler to recover from low and degraded tissue samples, giving it a distinct advantage over methods that are based on nuclear genomes for identifying species (Girish et al. 2004). The primer used for screening different mitochondrial genes to identify the fish species with their references has been mentioned in Table 2.1.

Table 2.1 List of primers for screening mitochondrial genes

Mitochondrial genes	Primer	References
Cytochrome oxidase I (COI)	F1: (5′-TCAACCAACCACAAAGACATTGGCAC-3′) R1: (5′-TAGACTTCTGGGTGGCCAAAGAATCA-3′)	Ward et al. (2005)
Cytochrome *b* (*Cyt-b*)	F: (5′-AAAAGCTTCCATCCAACATCTCAGCATGATGAA-3′) R: (5′-AAACTGCAGCCCCTCAGAATGATATTTGTCCTCA-3′)	Behera et al. (2015)
Dloop	F: (5′-AGCACCGGTCTTGTAAACCG-3′) R: (5′-GGGCTCATCTTAACATCTTCA-3′)	Cheng et al. (2012)
16S rRNA	F: (5′-CGCCTGTTTATCAAAAACAT-3′) R: (5′-CCGGTCTGAACTCAGATCACGT-3′)	Ivanova et al. (2007)
12S rRNA	F: (5′-ACTGGGATTAGATACCCC-3′) R: (5′-TAGAACAGGCTCCTCTAG-3′)	Miya et al. (2015)

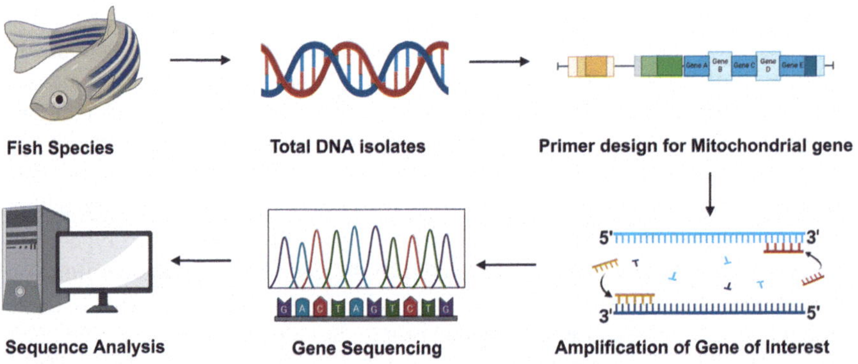

Fig. 2.1 Visual representation of identification fish species by using mitochondrial genes

In general, studies of vertebrate species have revealed that mitochondrial DNA acquires sequence divergence faster than nuclear DNA (Brown 1985). Mitochondrial DNA research has been effective in revealing connections among closely related species because of its high pace of evolution and it is known that distinct segment of the mtDNA change at different speeds (Chauhan et al. 2010; Meyer 1993). Furthermore, if gender-biased migration, selection, and introgression occur, the population and phylogenetic patterns obtained from mitochondrial DNA data may not provide a comprehensive picture of the nuclear genome (Birky et al. 1989; Chow and Kishino 1995). Non-coding portions, such as the D-loop of 1 kb length, have higher amounts of variation than coding sequences (Brown et al. 1993) whereas one of the slowest evolving genes in the mitochondrial genome is the 16S rRNA gene (Mitochondrial et al. 2022). DNA fragments of around 600 bp from the mitochondrial gene (COI) cytochrome oxidase I have been anticipated as a low-cost efficient and quick method for categorizing all biodiversity (Bingpeng et al. 2018; Donahue 2022; Hebert et al. 2004; Ward et al. 2009). Detailed information about mitochondrial genes and their function are described in Fig. 2.1.

2.2 Mitochondrial Genes and Their Structure

2.2.1 Cytochrome Oxidase I (COI)

A standardized 658 bp segment of the mitochondrial COI gene was anticipated in 2003 as a universal marker to be used as a "DNA barcode" tagging any taxon in the animal kingdom for species identification (Pentinsaari et al. 2016). The application of universal primers in PCR (polymerase chain reaction) to amplify and sequence the COI gene is known as barcoding (Chauhan et al. 2010). DNA barcoding, as it is known in molecular taxonomy utilizing the COI sequence, is not simply any DNA sequence but a stringently standardized sequence of a minimum length and quality from an agreed-upon gene (Costa and Carvalho 2007). In fact, the EMBL (European Molecular Biology Laboratory), NCBI (National Center for Biotechnology Information), and DDBJ (DNA Data Bank of Japan) have already decided that only COI sequences that adhere to these tight requirements will be accepted as DNA barcodes (Persis et al. 2009). The mitochondrial COI gene is a fundamental global fish bio-identification system component. Previous research has shown that characteristics such as sequence divergence across fish species are uncommon. Additionally, there is a sufficient level of sequence divergence to differentiate between related species (Hebert et al. 2003a, b; Karcz et al. 2012; Naseem and Tahir 2016). In the proton pumping process across the inner mitochondrial membrane, COX protein is involved in reducing O_2 to H_2O to form proton channels in the electron transport chain (Thomas et al. 1993).

The 3D structure of cytochrome *c* oxidase in a fish species *Schizothorax richardsonii* revealed that the tertiary structure of the COI gene is a helical structure linked by loops, forming a compact complex held together by two disulfide bonds (Chandra et al. 2015). The study on zebrafish revealed that the zebrafish *Cox-1* and *Cox-2* are identical to two mammalian Cox-related sequences which were reported previously (Grosser et al. 2002). The Cox-2b sequence in zebrafish comprises of 18 amino acid sequences in the C-terminal that are conserved in Cox-2 and absent in the Cox-1 protein of mammals (Herschman et al. 2003). The cDNA sequencing and comparison with the database genomic structure of Cox revealed that the zebrafish intron/exon segment of the zebrafish Cox-1 gene has a similar nucleotide arrangement to that of mouse Cox-1. In contrast, the comparison between Cox-2 sequences in murine and Cox-2a in Zebrafish results in missing one less exon as well as one intron. The Cox-1 gene in humans and mice is approximately 22–25 kb in length whereas in zebrafish the length is greater than 20 kb (Herschman et al. 2003). In contrast, the zebrafish Cox-2a and Cox-2b, Cox-2 genes are both quite short, measuring 5.5 and 7.3 kb in length, respectively (Ishikawa et al. 2007). A report based on data analysis by (Järving et al. 2004) advocates that the puffer fish, fugu (*Takifugu rubripes*) has three Cox-related sequences where two of the fugu cox genes are identical to human Cox-1, whereas the third fugu gene is more identical to human Cox-2 gene. Previous studies show that following the evolution of teleosts from tetrapods, the Cox-1 and Cox-2 genes were likely repeated; hence, Japanese killifish (*Oryzias latipes*) and puffer fishes (*Takifugu rubripes*) have two Cox-1 genes, while

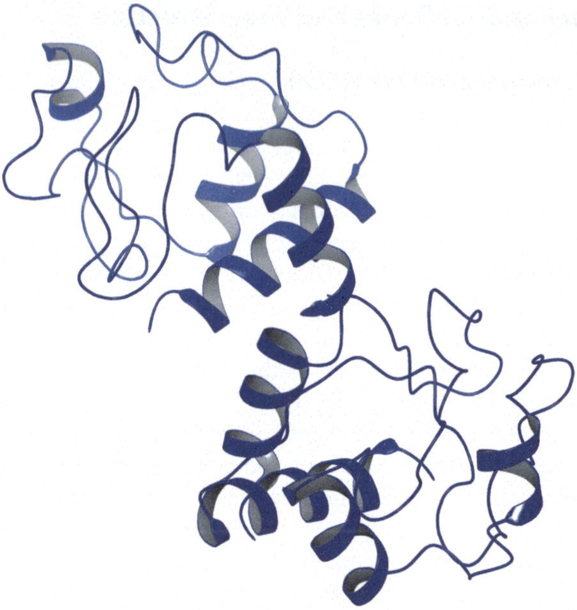

Fig. 2.2 3D protein structure of *cyt-c* gene of Tuna fish

rainbow trout and zebrafish kept two Cox-2 genes (Amores et al. 1998; Jatllon et al. 2004; Woods et al. 2005). The 3D protein structure of *cyt-c* gene of tuna fish (PDB ID 1 I55) is screened in Fig. 2.2.

2.2.2 Cytochrome-*b* Gene (*Cyt-b*)

In the mitochondrial genome, the *cyt-b* is a functional gene located between tRNA-^{Glu} and tRNA^{Thr} (Southern et al. 1988). The interaction of the *cyt b* gene with genomic DNA generates a complex enzyme cytochrome *c* oxidoreductase that is involved in oxidative phosphorylation (Leonard and Schapira 2000). *Cyt-b* is a protein present in the mitochondria of eukaryotic cells which is a component of the electron transport chain and serves as the main subunit of the cytochrome b6f and bc1complex. The *Cyt-b* gene has both slowly and quickly evolving codon locations, as well as more conservative and variable sections or domains overall. As a result, this gene has been used to study the diversity of species (Farias et al. 2014). In few prokaryotes and eukaryotes, *cyt-b* is a part of the bc1 complex (ubiquinol-cytochrome *c* reductase) that is essential for cellular respiration (Berry et al. 2003). These complexes are involved to create a proton-motive force in propelling of protons which is later used for ATP generation. The role of *cyt-b* is to bind covalently with two heme known as b562 and b566. The respiratory chain complex is made up of three subunits: *cyt b*, *cyt c1*, and one heme c bonded covalently and a single cluster of a possible Rieske iron sulfur protein (Ebenezer et al. 2015). The examination of the

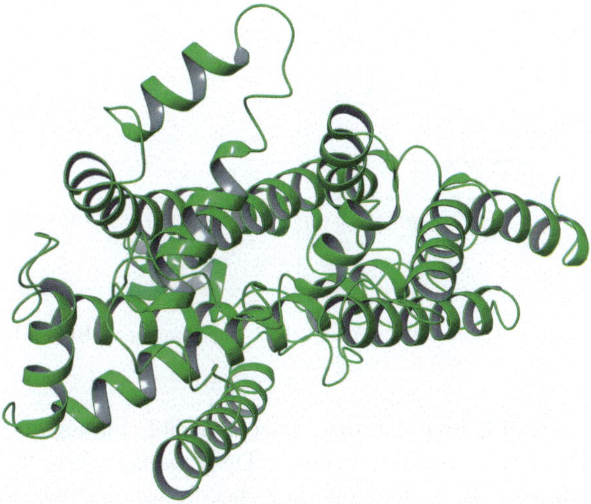

Fig. 2.3 3D structure of *cyt-b* gene isolated from *Danio rario*

secondary structure of the *cyt-b* protein of cold-water catfish *Glyptothorax ngapang* suggests that the most residue lies in the α-helix region (45.82%), the random coil region contains (31.90%), the extended region contains (17.93%), and the remaining 4.30% lies in β-turn. The instability index (II) of the protein is 32.14, isoelectric point (PI) is 7.36 (Singh et al. 2012). Similarly, cytochrome *b* of *carangoides equula* revealed the amino acid length was 380 having a size of 42.4 kDa and the PI value is 7.80. The 3D structure of *cyt-b* resembled the crystal structure of the cytochrome BC1 receptor (Ebenezer et al. 2015). Findings from analyses of mitochondrial *Cyt b* suggest that the most widely used DNA marker for fishes show that intraspecific variations across fishes are often less than 1%, however interspecific differences are typically more than 10% (Li et al. 2018). The alphafold 3D structure of *cyt-b* gene of *Danio rario* (Uniprot: Q9MIX8) is screened in Fig. 2.3.

2.2.3 rRNA Gene

One of the most adaptable sequences for phylogenetic analyses is the ribosomal gene and its accompanying spacers (Yan et al. 2013). By screening a range of species diversity within the target clade and increasing recognition of endangered species that support important and vulnerable functions, eDNA monitoring delivers a general resolution applicable to a variety of ecosystems (e.g., freshwater, brackish, and marine) for management and conservation applications (Mouillot et al. 2013). The effectiveness of a universal marker DNA for identification of species is dependent not only on sections of the selected gene targeted for PCR amplification but also on the sequence amplified between the forward and reverse primers being sufficiently differentiated between species (Karlsson and Holmlund 2007; Vences et al.

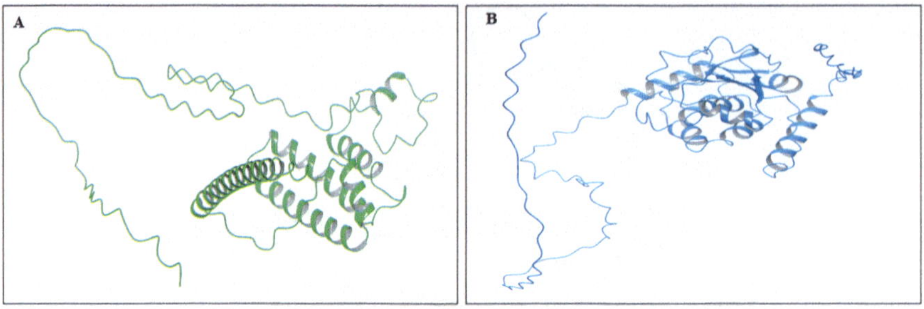

Fig. 2.4 (**a**) 3D structure of alpha fold 28S rRNA (RT15) gene segment. (**b**) 18S rRNA (zgc:92086) from zebrafish

2005). Since it is well known that there is considerable intra-species variation in mitochondrial DNA, it is crucial to choose a DNA marker where the inter-species variance is significantly larger than the intra-specific variance when unambiguous identifications must be achieved (Gharrett et al. 2001). The mitochondrial genes 18S and 12S rRNA are generally conserved, evolving at a slower rate than the mitochondrial genome (Aranishi et al. 2005). The mostly conserved sequences of these loci can be used as primer targeting sites, whereas mutations in variable areas have been shown to render both segments of DNA appropriate for species discrimination (Balitzki-Korte et al. 2005). The 18S rRNA gene has been demonstrated to be a more used target than the COI gene for fish species identification (Cawthorn et al. 2012; Vences et al. 2005). The study reported by Zhang et al. in 2020 revealed on average, primers targeting the 12S rRNA gene in mitochondria identified more diverse species of fish variety than those targeting the COI or 18S rRNA genes, while targeting the *cyt-b* gene amplified the fewest fish species in vitro. It also revealed that in terms of target gene diversity, 12S primers outperformed other primers in 53 amplified fish species (Zhang et al. 2020). The 28S rRNA is regarded as one of the fundamental elements of all eukaryotes and serves as the large subunit (LSU) of eukaryotic ribosomes (Awasthi and Serajuddin 2016). The large subunit ribosomal DNA (28S rRNA) is a mosaic of several varied and conservative segments that are commonly utilized as phylogenetic markers (Shylla et al. 2013). The 28S rRNA region in eukaryotic mitochondria has 12 divergent segments that differ significantly in nucleotide composition and length between species (de Rijk et al. 1995). The alpha fold 3D structure of 28S rRNA gene protein S15 of 263 amino acids (Uniport: Q6DGL8) and 18S rRNA zgc:92086 (Uniport: Q6DHT1) of 297 amino acids isolated from *Danio rerio* length is represented in Fig. 2.4.

2.2.4 D Loop

The D-loop in mtDNA is the hypervariable region of the mitochondrial genome, it comprises of the initiation point for heavy-strand replication as well as the promoter for light and heavy-strand transcription. Despite its functional relevance, this region is thought to be the most rapidly evolving region of a molecule, accumulating base substitutions at a pace significantly quicker than that of single-copy nucleus DNA (Brown et al. 1986; Saccone et al. 1987). Consequently, it has been extensively used as a biomarker to determine the phylogeographic history of a species (Avise 2000; Brown et al. 1993). The D-loop sequences are one of the most frequently utilized mitochondrial DNA sequences for characterizing the evolutionary significance of close relatives and subspecies because they are non-coding mitochondrial DNA (mtDNA) regions with a high rate of evolution and no significance of recombination (Murgia et al. 2002). Currently, D-loop sequences are commonly employed in genetic analysis for aquaculture species, particularly relating to genetic structure, phylogeny, molecular differentiation, genetic differentiation, and species validation (Brown and Thorgaard 2002; Ekerette et al. 2017; Tang et al. 2007; Yokoyama and Goto 2002). The characterization of fish species based on D-loop has already been reported in various fish like *Gymnocypris dobula, Carasobarbus luteus, Culter alburnus,* and *Clarias magur* (Chan et al. 2016; Parmaksiz 2019; Sahoo et al. 2020; Sun et al. 2002).

2.3 Conclusion

The innovation by using molecular biology techniques is increasing by utilizing mitochondrial genes as molecular markers to provide useful information for the management of fisheries and aquaculture. Most importantly, however, the new markers have stimulated new applications, such as the identification of the location of genes that regulate an organism's appearance and comprehension of how they express their effects in large populations of male and female fishes. Ultimately the choice of mitochondrial DNA as a tool for monitoring the diversity of Pisces should be guided by the aims of the study of environmental variable(s). The researcher should address the more traditional question related to characterizing fish diversity by using the mitochondrial marker genes as an efficient tool.

References

Alam MT, Petit RA, Read TD, Dove ADM (2014) The complete mitochondrial genome sequence of the world's largest fish, the whale shark (Rhincodon typus), and its comparison with those of related shark species. Gene 539:44–49

Amores A, Force A, Yan YL, Joly L, Amemiya C, Fritz A, Ho RK, Langeland J, Prince V, Wang YL, Westerfield M, Ekker M, Postlethwait JH (1998) Zebrafish hox clusters and vertebrate genome evolution. Science 1979(282):1711–1714

Anderson S, Bankier AT, Barrell BG, de Bruijn MHL, Coulson AR, Drouin J, Eperon IC, Nierlich DP, Roe BA, Sanger F, Schreier PH, Smith AJH, Staden R, Young IG (1981) Sequence and organization of the human mitochondrial genome. Nature 290(5806):457–465

Aranishi F, Okimoto T, Izumi S (2005) Identification of gadoid species (Pisces, Gadidae) by PCR-RFLP analysis. J Appl Genet 46:69–73

Attardi G (1988) Biogenesis of mitochondria. Annu Rev Cell Biol 4:289

Avise JC (2000) Phylogeography: the history and formation of species. Harvard University Press, Cambridge

Awasthi M, Serajuddin M (2016) Molecular phylogeny of four Gouramis based on divergent domain D11 of 28S rRNA gene. Adv Biores 10(6):351–353

Balitzki-Korte B, Anslinger K, Bartsch C, Rolf B (2005) Species identification by means of pyrosequencing the mitochondrial 12S rRNA gene. Int J Legal Med 119(5):291–294

Behera B, Singh N, Paria P, Sahoo AK, Panda D, Meena DK, Das P, Pakrashi S, Biswas DK, Sharma AP (2015) Population genetic structure of Indian shad, Tenualosa ilisha inferred from variation in mitochondrial DNA sequences. J Environ Biol 36(5):1193–1197

Berry EA, Guergova-Kuras M, Huang LS, Crofts AR (2003) Structure and function of cytochrome bc complexes. Annu Rev Biochem 69:1005–1075

Bingpeng X, Heshan L, Zhilan Z, Chunguang W, Yanguo W, Jianjun W (2018) DNA barcoding for identification of fish species in the Taiwan Strait. PLoS One 13:e0198109

Birky CW, Fuerst P, Maruyama T (1989) Organelle gene diversity under migration, mutation, and drift: equilibrium expectations, approach to equilibrium, effects of heteroplasmic cells, and comparison to nuclear genes. Genetics 121:613–627

Brown WM (1985) The mitochondrial genome of animals. In: MacIntyre RJ (ed) Molecular evolutionary genetics. Plenum Press, New York, pp 95–130

Brown KH, Thorgaard GH (2002) Mitochondrial and nuclear inheritance in an androgenetic line of rainbow trout, Oncorhynchus mykiss. Aquaculture 204:323–335

Brown GG, Gadaleta G, Pepe G, Saccone C, Sbisà E (1986) Structural conservation and variation in the D-loop-containing region of vertebrate mitochondrial DNA. J Mol Biol 192:503–511

Brown JR, Beckenbach AT, Smith MJ (1993) Intraspecific DNA sequence variation of the mitochondrial control region of white sturgeon (Acipenser transmontanus). Mol Biol Evol 10:326–341

Castro Paz FP, Batista JDS, Porto JIR (2014) DNA barcodes of Rosy Tetras and allied species (Characiformes: Characidae: Hyphessobrycon) from the Brazilian Amazon Basin. PLoS One 9:e98603

Cawthorn DM, Steinman HA, Witthuhn RC (2012) Evaluation of the 16S and 12S rRNA genes as universal markers for the identification of commercial fish species in South Africa. Gene 491:40–48

Chan J, Li W, Hu X, Liu Y, Xu Q (2016) Genetic diversity and population structure analysis of Qinghai-Tibetan plateau schizothoracine fish (Gymnocypris dobula) based on mtDNA D-loop sequences. Biochem Syst Ecol 69:152–160

Chandra S, Jain S, Mir JI, Singh BK (2015) Insilico characterization of cytochrome c oxidase I protein of Himalayan Snow Trout, Schizothorax richardsonii: 3-D structure prediction. World J Fish Mar Sci:87–97

Chauhan T, Rajiv K, Chauhan T, Rajiv K (2010) Molecular markers and their applications in fisheries and aquaculture. Adv Biosci Biotechnol 1:281–291

Cheng YZ, Xu TJ, Jin XX, Tang D, Wei T, Sun YY, Meng FQ, Shi G, Wang RX (2012) Universal primers for amplification of the complete mitochondrial control region in marine fish species. Mol Biol 46(5):727–730

Chow S, Kishino H (1995) Phylogenetic relationships between tuna species of the genus Thunnus (Scombridae: Teleostei): inconsistent implications from morphology, nuclear and mitochondrial genomes. J Mol Evol 41:741–748

Costa FO, Carvalho GR (2007) The barcode of life initiative: synopsis and prospective societal impacts of DNA barcoding of fish. Genom Soc Policy 3(2):1–12

de Rijk P, van de Peer Y, van den Broeck I, de Wachter R (1995) Evolution according to large ribosomal subunit RNA. J Mol Evol 41(3):366–375

Donahue MZ (2022) Collections after retirement: addressing the uncertain fate of a lifetime's work. Bioscience 72:718–724

Ebenezer NS, Maheswari U, Priyakumari J (2015) Homology modelling of cytochrome B from Carangoides equula. Int J Sci Res 6:2319–7064

Ekerette EE, Ikpeme EV, Udensi OU, Ozoje MO, Etukudo OM, Umoyen AJ, Durosaro SO, Wheto M, Ekerette EE, Ikpeme EV, Udensi OU, Ozoje MO, Etukudo OM, Umoyen AJ, Durosaro SO, Wheto M (2017) Phylogenetics and molecular divergence of tilapia fish (Oreochromis species) using mitochondrial D-loop and cytochrome b regions. Am J Mol Biol 8:39–57

Faith DP (1994) Genetic diversity and taxonomic priorities for conservation. Biol Conserv 68:69–74

Farias IP, Ortí G, Sampaio I, Schneider H, Meyer A (2014) The cytochrome b gene as a phylogenetic marker: the limits of resolution for analyzing relationships among cichlid fishes. J Mol Evol 53(2):89–103

Gharrett AJ, Gray AK, Heifetz J (2001) Identification of rockfish (Sebastes spp.) by restriction site analysis of the mitochondrial ND-3/ND-4 and 12S/16S rRNA gene regions. Fish Bull 99:49–49

Girish PS, Anjaneyulu ASR, Viswas KN, Anand M, Rajkumar N, Shivakumar BM, Bhaskar S (2004) Sequence analysis of mitochondrial 12S rRNA gene can identify meat species. Meat Sci 66:551–556

Gregory TR (2009) Understanding natural selection: essential concepts and common misconceptions. Evol Educ Outreach 2:156–175

Grosser T, Yusuff S, Cheskis E, Pack MA, FitzGerald GA (2002) Developmental expression of functional cyclooxygenases in zebrafish. Proc Natl Acad Sci 99:8418–8423

Hebert PDN, Cywinska A, Ball SL, DeWaard JR (2003a) Biological identifications through DNA barcodes. Proc Biol Sci 270:313–321

Hebert PDN, Ratnasingham S, DeWaard JR (2003b) Barcoding animal life: cytochrome c oxidase subunit 1 divergences among closely related species. Proc R Soc B Biol Sci 270:S96

Hebert PDN, Stoeckle MY, Zemlak TS, Francis CM (2004) Identification of birds through DNA barcodes. PLoS Biol 2:e312

Herschman HR, Talley JJ, DuBois R (2003) Cyclooxygenase 2 (COX-2) as a target for therapy and noninvasive imaging. Mol Imaging Biol 5:286–303

Ishikawa T-O, Griffin KJP, Banerjee U, Herschman HR (2007) The zebrafish genome contains two inducible, functional cyclooxygenase-2 genes. Biochem Biophys Res Commun 352:181

Ivanova NV, Zemlak TS, Hanner RH, Hebert PDN (2007) Universal primer cocktails for fish DNA barcoding. Mol Ecol Notes 7:544–548

Järving R, Järving I, Kurg R, Brash AR, Samel N (2004) On the evolutionary origin of cyclooxygenase (COX) isozymes. J Biol Chem 279:13624–13633

Jatllon O, Aury JM, Brunet F, Petit JL, Stange-Thomann N, Maucell E, Bouneau L, Fischer C, Ozouf-Costaz C, Bernot A, Nicaud S, Jaffe D, Fisher S, Lutfalla G, Dossat C, Segurens B, Dasilva C, Salanoubat M, Levy M, Houdet N, Castellano S, Anthouard V, Jubin C, Castelli V, Katinka M, Vacherie B, Blémont C, Skalli Z, Cattolico L, Poulain J, de Berardinis V, Cruaud C, Dupart S, Brottler P, Coutanceau JP, Gouzy J, Parra G, Lardier G, Chapple C, McKernan KJ, McEwan P, Bosak S, Kellis M, Volff JN, Gulgó R, Zody MC, Mesirov J, Lindblad-Toh K, Birren B, Nusbaum C, Kahn D, Robinson-Rechavi M, Laudet V, Schachter V, Quétler F, Saurin W, Scarpeill C, Wincker P, Lander ES, Weissenbach J, Roest Crollius H (2004) Genome duplication in the teleost fish Tetraodon nigroviridis reveals the early vertebrate proto-karyotype. Nature 431(7011):946–957

Kali Mahapatra B, Kumar Sarkar U (2015) Sustainable management of floodplain wetlands for enhanced fishery and livelihood view project diversification of enclosure culture in reservoirs and wetlands. View project. J Biodivers Biopros Dev 2:1

Karcz J, Bernas T, Nowak A, Talik E, Woznica A (2012) Application of lyophilization to prepare the nitrifying bacterial biofilm for imaging with scanning electron microscopy. Scanning 34:26–36

Karlsson AO, Holmlund G (2007) Identification of mammal species using species-specific DNA pyrosequencing. Forensic Sci Int 173:16–20

Ladoukakis ED, Zouros E (2017) Evolution and inheritance of animal mitochondrial DNA: rules and exceptions. J Biol Res (Greece) 24:1–7

Leonard JV, Schapira AHV (2000) Mitochondrial respiratory chain disorders I: mitochondrial DNA defects. Lancet 355:299–304

Li X, Shen X, Chen X, Xiang D, Murphy RW, Shen Y (2018) Detection of potential problematic Cytb gene sequences of fishes in GenBank. Front Genet 9:30

Liu Z, Karsi A, Dunham RA (1999) Development of polymorphic EST markers suitable for genetic linkage mapping of catfish. Mar Biotechnol 1(5):437–447

Merheb M, Matar R, Hodeify R, Siddiqui SS, Vazhappilly CG, Marton J, Azharuddin S, Zouabi HAL (2019) Mitochondrial DNA, a powerful tool to decipher ancient human civilization from domestication to music, and to uncover historical murder cases. Cells 8:433

Meyer A (1993) Evolution of mitochondrial DNA in fishes. In: Mochachka PW, Mommsen TP (eds) Biochemistry and molecular biology of fishes. Elsevier, Amsterdam, pp 1–38

Mitochondrial S, Solari P, Sollai G, Massimo Crnjar R, Giglio A, Wink M, Ye W, Wang J, Zhao X, Liu H, Zhu S (2022) Mitochondrial genomes of two Lycosa spiders (Araneae, Lycosidae): genome description and phylogenetic implications. Diversity 14:538

Miya M, Sato Y, Fukunaga T, Sado T, Poulsen JY, Sato K, Minamoto T, Yamamoto S, Yamanaka H, Araki H, Kondoh M, Iwasaki W (2015) MiFish, a set of universal PCR primers for metabarcoding environmental DNA from fishes: detection of more than 230 subtropical marine species. R Soc Open Sci 2:150088

Mouillot D, Bellwood DR, Baraloto C, Chave J, Galzin R, Harmelin-Vivien M, Kulbicki M, Lavergne S, Lavorel S, Mouquet N, Paine CET, Renaud J, Thuiller W (2013) Rare species support vulnerable functions in high-diversity ecosystems. PLoS Biol 11:e1001569

Murgia R, Tola G, Archer SN, Vallerga S, Hirano J (2002) Genetic identification of grey mullet species (Mugilidae) by analysis of mitochondrial DNA sequence: application to identify the origin of processed ovary products (bottarga). Mar Biotechnol 4(2):119–126

Naseem S, Tahir HM (2016) Use of mitochondrial COI gene for the identification of family Salticidae and Lycosidae of spiders. Mitochondrial DNA A DNA Mapp Seq Anal 29:96–101

Parmaksiz A (2019) Population genetic diversity of yellow barbell (Carasobarbus luteus) from Kueik, Euphrates and Tigris Rivers based on mitochondrial DNA D-loop sequences. Turk J Fish Aquat Sci 20:79–86

Pentinsaari M, Salmela H, Mutanen M, Roslin T (2016) Molecular evolution of a widely-adopted taxonomic marker (COI) across the animal tree of life. Sci Rep 6:35275

Persis M, Reddy ACS, Rao LM, Khedkar GD, Ravinder K, Nasruddin K (2009) COI (cytochrome oxidase-I) sequence-based studies of Carangid fishes from Kakinada coast, India. Mol Biol Rep 36:1733–1740

Pirger Z, Rácz B, Kiss T (2009) Dopamine-induced programmed cell death is associated with cytochrome c release and caspase-3 activation in snail salivary gland cells. Biol Cell 101:105–116

Preziosi P (2007) Drug development. In: Comprehensive medicinal chemistry, vol II. Elsevier, Amsterdam, pp 173–202

Ruan H, Li M, Li Z, Huang J, Chen W, Sun J, Liu L, Zou K (2020) Comparative analysis of complete mitochondrial genomes of three Gerres fishes (Perciformes: Gerreidae) and primary exploration of their evolution history. Int J Mol Sci 21:1874

Saccone C, Attimonelli M, Sbisà E (1987) Structural elements highly preserved during the evolution of the D-loop-containing region in vertebrate mitochondrial DNA. J Mol Evol 26(3):205–211

Sahoo L, Barat A, Sahoo SK, Sahoo B, Das G, Das P, Sundaray JK, Swain SK (2020) Genetic diversity and population structure of endangered Indian catfish, Clarias magur as revealed by mtDNA D-loop marker. Turk J Fish Aquat Sci 21:9–18

Shirasawa K, Hirakawa H (2013) DNA marker applications to molecular genetics and genomics in tomato. Breed Sci 63:21

Shylla JA, Ghatani S, Tandon V (2013) Utility of divergent domains of 28S ribosomal RNA in species discrimination of paramphistomes (Trematoda: Digenea: Paramphistomoidea). Parasitol Res 112:4239–4253

Singh M, Gupta A, Lakra WS (2012) In silico 3-D structure prediction of cytochrome b protein of sisorid catfish Glyptothorax ngapang. Indian J Biotechnol 11:156–162

Southern ŠO, Southern PJ, Dizon AE (1988) Molecular characterization of a cloned dolphin mitochondrial genome. J Mol Evol 28:32–42
Sun YH, Wang W, Liu SY, He SP, Shao XL, Xie ZX, Deng FJ, Liu Y, Tong JG, Wu QJ (2002) [Genetic diversity analysis of mitochondrial D-loop region of Chinese sucker (Myxocyprinus asiaticus)]. Yi Chuan Xue Bao 29:787–790
Sun CH, Liu HY, Xu N, Zhang XL, Zhang Q, Han BP (2021) Mitochondrial genome structures and phylogenetic analyses of two tropical Characidae fishes. Front Genet 12:80
Tang W, Hu X, Yang J (2007) Species validities of Coilia brachygnathus and C. nasus taihuensis based on sequence variations of complete mtDNA control region. Biodivers Sci 15:224
Tanya C, Kumar R (2010) Molecular markers and their applications in fisheries and aquaculture. Adv Biosci Biotechnol 2010:281–291
Thomas JW, Gennis RB, Puustinen A, Wikström M, Alben JO (1993) Substitution of asparagine for Aspartate-135 in subunit I of the cytochrome bo ubiquinol oxidase of Escherichia coli eliminates proton-pumping activity. Biochemistry 32:10923–10928
Vences M, Thomas M, van der Meijden A, Chiari Y, Vieites DR (2005) Comparative performance of the 16S rRNA gene in DNA barcoding of amphibians. Front Zool 2:1–12
Wallace DC (1999) Mitochondrial diseases in man and mouse. Science 283(5407):1482–1488
Ward RD, Zemlak TS, Innes BH, Last PR, Hebert PDN (2005) DNA barcoding Australia's fish species. Philos Trans R Soc B Biol Sci 360:1847–1857
Ward RD, Hanner R, Hebert PDN (2009) The campaign to DNA barcodes all fishes, FISH-BOL. J Fish Biol 74:329–356
Woods IG, Wilson C, Friedlander B, Chang P, Reyes DK, Nix R, Kelly PD, Chu F, Postlethwait JH, Talbot WS (2005) The zebrafish gene map defines ancestral vertebrate chromosomes. Genome Res 15:1307–1314
Wu Y, Wang X, Liu S, Luo H, Lin Q (2019) Population genetic structure and phylogenetic analysis of gray's pipefish, Halicampus grayi in the South China Sea. Genes Genomics 42(2):155–164
Yan H, Lou Z, Li L, Ni X, Guo A, Li H, Zheng Y, Dyachenko V, Jia W (2013) The nuclear 18S ribosomal RNA gene as a source of phylogenetic information in the genus Taenia. Parasitol Res 112:1343–1347
Yang W, Kang X, Yang Q, Lin Y, Fang M (2013) Review on the development of genotyping methods for assessing farm animal diversity. J Anim Sci Biotechnol 4:1–6
Yang L, Tan Z, Wang D, Xue L, Guan MX, Huang T, Li R (2014) Species identification through mitochondrial rRNA genetic analysis. Sci Rep 4:4089
Yokoyama R, Goto A (2002) Phylogeography of a freshwater sculpin, Cottus nozawae, from the northeastern part of Honshu Island, Japan. Ichthyol Res 49(2):147–155
Zhang S, Zhao J, Yao M (2020) A comprehensive and comparative evaluation of primers for metabarcoding eDNA from fish. Methods Ecol Evol 11:1609–1625

Role of Epigenetics in Fisheries and Aquaculture

3

Kiran D. Rasal, Prachi Asgolkar, Siba Shinde, Diganta Dey, and Jitendra Kumar Sundaray

Abstract

High-throughput sequencing approaches are immensely used in fisheries research for solving biological questions. This has led to discovering of molecular markers associated with production or performance traits in cultured fish. Recently, interest in epigenetics has surged as it has become clearer that epigenetic mechanisms can provide a measurable link between environment and phenotype. Epigenetics includes genome function alteration without changing DNA sequence, including DNA methylation (cytosine), histone modifications (e.g., lysine and arginine methylation), non-coding RNAs, etc. DNA methylation at cytosine, which forms 5-methylcytosine (5mC), is vital in regulating gene expression and influences various biological processes. Recently developed computational tools allow rapid discovery of methylation sites at a genome-wide scale and demonstrate that environmental factors such as nutrition, stress, and exposure to compounds affect epigenetic gene regulation, which alters their phenotype. In this book chapter, we discussed the importance of epigenetics in aquaculture, methods to detect and the status of aquaculture species.

K. D. Rasal
Fish Genetics and Biotechnology Division, ICAR-Central Institute of Freshwater Aquaculture, Bhubaneswar, Odisha, India

Fish Genetics and Biotechnology, ICAR-Central Institute of Fisheries Education, Mumbai, Maharashtra, India

P. Asgolkar · S. Shinde · D. Dey
Fish Genetics and Biotechnology, ICAR-Central Institute of Fisheries Education, Mumbai, Maharashtra, India

J. K. Sundaray (✉)
Fish Genetics and Biotechnology Division, ICAR-Central Institute of Freshwater Aquaculture, Bhubaneswar, Odisha, India

B. K. Behera (ed.), *Current Trends in Fisheries Biotechnology*,
https://doi.org/10.1007/978-981-97-3165-7_3

Keywords

DNA methylation · Next-generation sequencing · Epigenetics

3.1 Introduction

Over the past 10 years, there has been notable advancement in fish genomics, leading to the availability of genome sequences for several aquaculture species in public databases. Thus, availability of those genome sequences opens an avenue for molecular marker discovery associated with production and performance traits (Abdelrahman et al. 2017; Canário 2019; Lu and Luo 2020; Yuan et al. 2018). Studies suggest that important traits such as growth, disease resistance, FCR, stress tolerance and many others in aquaculture species are mainly controlled by genetic and epigenetic variations. The understanding of epigenetics regulation in fish species is often ignored. The genome sequences of fish species allowed the mapping of SNPs or methylated sites linked with production traits. The advancement in sequencing technology helps decipher epigenomic variation at a genome-wide scale.

Epigenetics deals with the alteration of gene function without change of underlying genome sequence possibly due to environmental stimuli. Epigenetic regulation mainly includes DNA methylation, non-coding RNAs or small RNAs and histone modifications. DNA methylation in eukaryotes consists of cytosine methylation at 5mC, the main methylated form of cytosine (Moore et al. 2013; Zeng and Chen 2019). This type of methylation exerts their function at the transcriptional level via either activation or repression of gene. Studies implicated that DNA methylation could affect the organism's growth, stress, reproduction, and many other developmental processes. Stable methylation is observed in somatic cells, while sex-specific methylation is established in germ cells which includes global de-methylation and re-methylation (Cavalli and Heard 2019). Recent few studies revealed that epigenetic factors control the developmental biology of non-model aquaculture species (Kelley et al. 2021; Liu et al. 2022).

Although epigenomic research in fish species is still at an early stage, many researchers are attempting to understand epigenetic regulation associated with important traits. This book chapter reviews progress in epigenetic research in fisheries and candidate aquaculture species, possible perspectives, and challenges.

3.2 Types of Epigenetic Modifications

Epigenetics is the investigation of heritable modifications in gene expression that do not involve changes to the DNA sequence. Environmental cues are now recognized to trigger animal phenotypic changes through epigenetic mechanisms that can regulate gene expression at a molecular scale without altering the DNA nucleotide sequence (Jaenisch and Bird 2003). Epigenetic controls encompass DNA methylation, non-coding RNA activity, histone changes, and chromatin remodelling. The modifications that happen during epigenetics are heritable.

3.2.1 DNA Methylation

In addition, with other regulators, DNA methylation is now widely recognized as an essential epigenetic mechanism affecting gene activity (Jin et al. 2011). DNA methylation is an epigenetic modification that involves the addition of a methyl group to the cytosine base at the C5 position, resulting in the formation of 5-methylcytosine specifically in CpG dinucleotides. DNA methylation regulates gene expression by attracting proteins that suppress genes or blocking the attachment of transcription factors to DNA. Various cells develop a unique and stable DNA methylation pattern through a dynamic process that includes de novo DNA methylation and demethylation. This pattern controls gene transcription specific to different tissues (Moore et al. 2013).

A class of enzymes known as DNA methyltransferases (Dnmts) is responsible for catalysing the methylation of DNA. Methyl groups are transferred from *S*-adenyl methionine (SAM) to the fifth carbon of a cytosine residue, resulting in the production of 5mC. Dnmt3a and Dnmt3b are responsible for creating a new methylation pattern on unmodified DNA, referred to as de novo Dnmt. On the other hand, Dnmt1 plays a role in DNA replication by transferring the DNA methylation pattern from the parent DNA strand to the newly synthesized daughter strand, ensuring semi-conservative replication of methylation (Jablonka and Raz 2009; Moore et al. 2013). All three Dnmts play a role in embryonic development. When cells reach the final phase of differentiation, Dnmt expression significantly decreases. Postmitotic cells exhibit a consistent DNA methylation pattern. The mature mammalian brain, however, still exhibits considerable amounts of postmitotic neurons' Dnmts expression, suggesting that Dnmts and DNA methylation may have a novel function in the brain (Feng et al. 2005; Morris and Monteggia 2014). Neurons react to their environment through depolarization patterns, transmitting information and encoding a response. Recent research has shown that alterations in gene expression after depolarization are linked to changes in the epigenetic profile, including variations in DNA methylation.

Therefore, this epigenetic mechanism is crucial for many biological processes, such as silencing retroviral elements, regulation of tissue-specific gene expression, genomic imprinting, stabilization of chromosomal structure, X chromosome inactivation, normal development, regulation and maintenance of gene activity patterns, DNA replication and repair. The majority of DNA methylation takes place at CpG sites, which are cytosines that come before guanine nucleotides

3.2.2 Non-coding RNAs

Most eukaryotic genomes are transcribed into non-coding RNAs (ncRNAs) that do not encode proteins. These ncRNAs play diverse regulatory roles in mammalian transcriptomes and are crucial for controlling gene expression through epigenetic mechanisms. These non-coding RNAs can control cytosine methylation and histone modifications that impact gene expression regulation in multicellular animals.

These molecules play crucial roles in transposon activity and silencing, X-chromosome inactivation, position effect variegation and paramutation. Evidence indicates that non-coding RNA plays a crucial part in the silencing process of autosomal chromosomes, as they are consistently transcribed from an imprinted region in mammalian genomes. This closely resembles the processes employed by the big non-coding RNA XIST during X-chromosome inactivation in female cells (Lee 2009). Evidence has connected an RNA component, mainly ncRNA, to gene regulation and chromatin remodelling. The connection between non-coding RNA and the process of heterochromatin suppression shows the importance of these transcripts in maintaining a typical chromatin structure. Short RNAs are increasingly recognized for their involvement in suppressing repetitive sequences in the genome, as demonstrated by Chu and Rana in 2007.

The zebrafish study on PiRNA demonstrated their involvement in silencing repetitive elements in vertebrates (Houwing et al. 2007). DNA methylation is responsible for inactivating one copy of the sexual chromosome in females, and the XIST ncRNA is a gene product thought to be involved in this process. A 17 kb long non-coding RNA called XIST can physically attach and form complexes with the chromatin surrounding one of the female X-chromosomes (Ng et al. 2007). The study on long antisense non-coding RNA (ncRNA) called HOTAIR revealed a new mechanism where the transcription of ncRNA controls the transcriptional silencing of a chromosomal region. HOTAIR is also involved in mediating epigenetic silencing of chromosomal domains by directing specific histone modifications (Rinn et al. 2007).

3.2.3 Histone Modifications and Chromatin Remodelling

Chromatin contains 8 subunit histones and DNA coiled, 146 bp of DNA coil around 1 histone, DNA is compacted and can fit into the nucleus. Chromatin can interchangeably adopt either of the two primary states. These are euchromatin and heterochromatin, respectively. The compact form of heterochromatin, which includes transcriptional machinery, is resistant to the binding of many different proteins. On the other hand, Euchromatin is a more relaxed type of chromatin susceptible to transcriptional processes and changes (Khalaf and Gräff 2016). Waddington first used the term "epigenetics" in 1942 to refer to how genes interact with the environment to create phenotypes. In response to environmental cues, the epigenome undergoes biochemical modifications that alter the structure of the chromatin. The word "chromatin remodelling" primarily refers to the ATP-dependent change of the genome caused by nucleosome-shifting enzymes like the SWI/SNF complex. The exposure of a particular gene's promoter to its transcriptional machinery is one biological process made possible by this mechanical alteration in genomic conformation (Narlikar et al. 2013). A mammal undergoes epigenetic reprogramming twice during its lifetime: during gametogenesis and the early stages of embryonic development. During spermatogenesis, there are global sex-specific alterations to the epigenome characterized by a sequence of events starting with DNA demethylation, then DNA methylation, and finally chromatin modifications. In germ cells, the epigenetic regulation of gene function follows particular protocols. Male and female gametes' development and the epigenetic state of their genomes show notable sex-specific differences. During gametogenesis, the male and female-specific epigenetic programs are reset to allow the gametes to combine at fertilization and form a totipotent zygote capable of generating any cell. For aberrant epigenetic alterations to be avoided from being passed on to the next generation, the fidelity of the resetting procedure is crucial. Offspring of parents with impaired epigenetic reprogramming and exposure to damaging environmental or chemical stimuli may have increased vulnerability to diseases due to their male and female germ cells (Morgan et al. 2005).

3.3 Methods to Study DNA Methylation and Histone Modification

Various established approaches are widely used to evaluate DNA methylation status at different levels and dimensions, reflecting the significant advancements in the field over the past few years.

Bisulphite sequencing, oxidative bisulphite sequencing, methylated DNA immunoprecipitation (MeDIP), reduced representation bisulphite sequencing (RRBS), and antibody-based ELISA method are some of the available methods for DNA methylome assessment.

One of the most significant epigenetic controls, histone modification, has been shown to have a regular impact on gene expression by affecting transcription. Histones are subjected to a variety of post-translational modifications. Histone modification can cause alteration in its electrostatic charge that affects binding to DNA or it can lead to the generation of protein recognition module binding sites. Mass spectrometry and genomics approaches like ChIP-chip and ChIP-seq can be used for analysing histone modification.

3.3.1 Bisulphite Sequencing

Bisulphite genomic sequencing facilitates identification of 5-methylcytosine at single base pair resolution therefore it is considered to be the standard method to assess DNA methylome. This method was first described by Frommer et al. in 1992. It is based on the principle that after treatment with sodium bisulphite the unmethylated cytosine undergoes deamination and gets converted into uracil while methylated cytosine does not undergo deamination and remains in its original form. Single-stranded DNA cytosines are converted to uracil residues, which are then recognized as thymine during PCR amplification and sequencing. However, 5 mCs remain as cytosines and can be differentiated from unmethylated cytosines (Li et al. 2021). As a core technique for analysing DNA methylation, bisulphite genomic sequencing has been widely applied in a variety of research and therapeutic situations. We have given flowchart of bisulphite sequencing in Fig. 3.1.

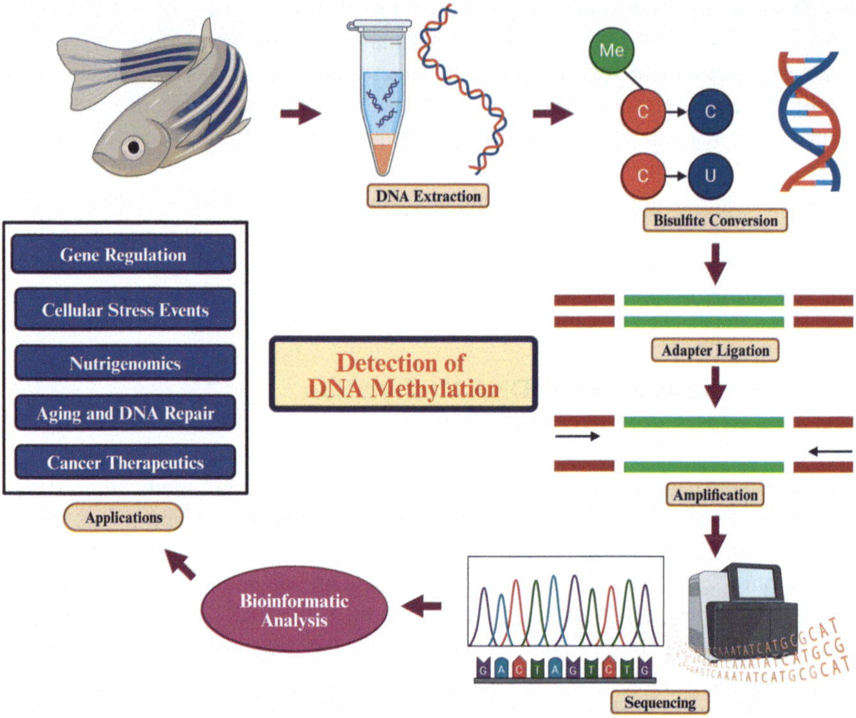

Fig. 3.1 Flowchart of bisulphite sequencing method using high-throughput sequencing

3.3.2 Methylated DNA Immunoprecipitation Sequencing (MeDIP)

Using an antibody against 5-methylcytosine, methylated DNA is immunoprecipitated, then it is sequenced. Comparing the immunoprecipitated DNA to the reference genome enables the identification of the methylation DNA segment. Methylated DNA immunoprecipitation (MeDIP) is a reliable method for isolating methylated DNA from a target material (Harrison and Parle-McDermott 2011).

3.3.3 Whole Genome Sequencing

With sodium bisulphite treatment followed by whole genome sequencing, an entire DNA methylome can be assessed at a single nucleotide resolution. This method was used to create an *A. thaliana* DNA methylation map (Cokus et al. 2008).

3.3.4 Mass Spectrometry

Mass spectrometry is a simple technique that utilizes differences in charge to mass ratio to separate ions. Differences in expected protein mass due to modification can be identified using mass spectrometry if the primary amino acid sequence is known. Hence, mass spectrometry is an excellent tool for identifying histone post-translational modifications (Freitas et al. 2004).

3.3.5 Chromatin Immunoprecipitation Technique (ChIP)

Chromatin immunoprecipitation (ChIP) is an effective method for studying histone modifications. The chromatin immunoprecipitation (ChIP) technique is the most widely used approach for determining and quantifying chromatin changes and interaction patterns. The ChIP assay employs antibodies targeting particular histone modification indicators, epigenetic modulators, and specific DNA fragments. Typically, ChIP is used to identify the relative abundance of specific protein or protein modification at a certain region in the genome. ChIP involves cross-linking DNA and protein, then fragmentation of the chromatin, immunoprecipitation with protein-specific antibody, cross-link reversal and DNA purification then evaluation by qPCR. Recent technological development like ChIP-on-ChIP and ChIP Sequencing can help us to get a view of genome-wide histone modifications (Singh and Szabó 2012).

3.4 Status of Epigenetics in Fish

Global aquaculture production is expanding quickly, and in order to meet the growing demand for fish, new technological advancements are urgently needed in a variety of fields, including disease control, breeding, feeds, and nutrition. Regarding its potential value in numerous agricultural production areas, epigenetics has received a lot of attention. It offers significant prospects for the aquaculture sectors, especially in areas where the environment is adaptable. Studying disease resistance and other economically important traits in commercially valuable species through epigenetic mechanisms and markers can provide significant opportunities and boost economic revenue in the aquaculture industry (Abdelrahman et al. 2017; Canário 2019). Several studies have demonstrated epigenetic effects in different animals resulting from different environments that promote beneficial adaptive physical characteristics. These effects are generally passed down over generations, making them particularly important in the context of production. The inheritance of epigenetic markers and phenotypes across generations highlights the importance of applying epigenetics in aquaculture to achieve greater environmental and economic sustainability. We have given a detailed epigenetic study undertaken in various fish species in Table 3.1.

3.4.1 Resistance to Disease and Stress

Some abiotic conditions have been linked in studies to altered epigenetic states or modifications in epigenetic markers that result in enhanced disease resistance and heightened stress tolerance, potentially yielding economic advantages (González-Recio 2011). Early-life stress can affect an aquatic animal's ability to cope with diseases and other stressors in adulthood. Early life temperature stress has been associated with alterations in Atlantic salmon's methylome and transcriptome patterns, affecting disease resistance and immune response (Moghadam et al. 2017; Uren Webster et al. 2018). Research has shown that brine shrimps exhibit heat shock-induced thermotolerance and resistance to pathogens (Norouzitallab et al. 2014). When given to the parental generation early in life, a plant-derived substance called phloroglucinol triggers the production of heat shock protein (hsp). This leads to transgenerational epigenetic inheritance resistance in offspring against Vibrio infections and thermal stress (Roy et al. 2019). The study indicated that the observed phenotype was influenced by heightened activity of innate immunity genes and epigenetic processes, including histone modifications (acetylation and methylation), DNA methylation, and RNA m6A methylation. Resistance levels rose in three consecutive generations. Furthermore, increasing evidence suggests that early immunological stressors could enhance immune responses in later life and transmit memory to the next generation. Research indicates that offspring of Pacific oyster parents treated with poly (I:C) exhibit enhanced defences against Ostreid herpes virus type I infection (Green et al. 2016). Oyster larvae from parents treated with poly (I:C) exhibited improved defence characteristics and higher levels of gene expression related to antiviral activity, indicating inheritable epigenetic changes. A study conducted with the invertebrate model Artemia found that the F1-F3 offspring of parents exposed to the pathogenic *Vibrio campbellii* at a young age exhibited a resistant characteristic compared to the progeny of control parents. This indicates potential epigenetic reprogramming (Norouzitallab et al. 2014). Moreover, disease resistance was associated with epigenetic reprogramming, including histone changes, m6A RNA methylation, and increased transcription of innate immunity genes. These discoveries have the potential to benefit the sustainability of aquaculture significantly.

3.4.2 Nutrigenomics

Early nutritional programming and its impact on significant phenotypic features later in life is essential in studying dietary epigenetics to enhance performance. Some research has indicated a connection between early feeding and epigenetics, which could benefit aquaculture. Later transcriptome investigations suggested that epigenetic mechanisms may have affected the observed response (Balasubramanian et al. 2016). Vitamin supplementation at the initial feeding stage led to changes in global methylation and histone modification 7 months after the supplementation was discontinued, despite the absence of visible physical reactions (Panserat et al. 2017). Studies in Senegalese sole (*Solea senegalensis*) showed that temperature can

Table 3.1 Methylation status in different fish species using NGS approaches

Species	Stress factor	Exposure conditions	Epigenetic mark studied	Methodology	Main epigenetic results	Phenotypic outcome	References
Dicentrarchus labrax (European sea bass)	Temperature	Exposure to higher temperature during the thermosensitive period	Differential methylation of gonadal *cyp19a* and *b-actin* promoters	Bisulphite sequencing (BS-Seq)	Increased methylation levels of the gonadal aromatase (*cyp19a*) gene promoter	Masculinization of females	Navarro-Martín et al. (2011)
Cynoglossus semilaevis (half-smooth tongue sole)	Temperature	Exposure to higher temperature during the sensitive developmental period in early life	DNA methylation of samples from gonadal tissue, methylation status of *dmrt1*	Bisulphite sequencing (BS-Seq)	Substantial methylation of genes involved in the sex determination network during sexual reversal; global inherited by the offspring	Masculinization of females	Shao et al. (2014)
Artemia (Brine shrimp)	Temperature	Nonlethal heat shock during early life stages	Global DNA methylation, histone H3 and H4 total acetylation	Restriction digestion and UHPLC-MS/MS for global DNA methylation; Western blot and immunofluorescence analysis for histone H4 and H3 acetylation, respectively	Altered levels of both global DNA methylations and increased total acetylation of histones H3 and H4	Tolerance to lethal heat stress and resistance against the pathogenic *Vibrio campbellii*; Phenotype inherited by three subsequent unstressed generations	Norouzitallab et al. (2014)

(continued)

Table 3.1 (continued)

Species	Stress factor	Exposure conditions	Epigenetic mark studied	Methodology	Main epigenetic results	Phenotypic outcome	References
Solea senegalensis (Senegalese sole)	Temperature	Larvae undergoing metamorphosis reared at different temperatures	Cytosine methylation of the *myog* gene promoter	Bisulphite sequencing (BS-seq)	Increased methylation of *myog* gene promoter in skeletal muscle, at lower temperature; increased *dnmt1* and *dnmt3s* expression levels	Increased thermal plasticity of myogenesis of larvae reared at higher temperatures	Campos et al. (2013)
Gadus morhua (Atlantic cod)	Photoperiod	Exposure to two different photoperiod regimes during early juvenile stages	DNA methyltransferases (*dnmts*) expression in fast muscle tissue	Semi-quantitative and quantitative real-time PCR (qPCR)	Increased expression of *dnmt1* and *dnmt3a* in fast muscle of Atlantic cod exposed to continuous illumination relative to fish reared under a natural photoperiod	Organisms reared under continuous light during the early juvenile stages presented larger size compared to fish reared at simulated natural photoperiod	Giannetto et al. (2013)
Salmo trutta (Brown trout)	Nutrition, salinity	Exposure of freshwater fish to seawater conditions, after salt-enriched diet intake	Cytosine methylation in CpG sites of samples from gill tissue	Methyl-sensitive amplification polymorphism (MSAP) techniques	Reversible and temporary salt-induced changes in global DNA methylation	Freshwater fish physiologically adapted to seawater	Morán et al. (2013)

Anguillla anguilla (European eel)	Metal exposure	Chronic exposure of immature organisms to cadmium	Global and site-specific CpG methylation in liver tissue	Enzyme-linked immunosorbent assay (ELISA) and methylation-sensitive arbitrarily primed PCR method (msAP-PCR)	Increased levels of global CpG methylation; hypermethylation of genes related to intracellular trafficking and phospholipid biosynthetic processes	–	Pierron et al. (2014)
Danio rerio (Zebrafish)	Temperature and metal exposure	Exposure to copper and heat stress during early embryogenesis	Global DNA methylation; stress-related gene expression	Colorimetric technique–Methyl Flash Methylated DNA Quantification Kit	No difference in global cytosine methylation, but mRNA expression of *dnmt3* genes was affected; increased mRNA expression of *mt2* and *hsp70* due to copper exposure; increased *hsp70* mRNA expression at higher temperature	High mortality, delayed embryo hatching and reduced hatching rate	Dorts et al. (2016)

(continued)

Table 3.1 (continued)

Species	Stress factor	Exposure conditions	Epigenetic mark studied	Methodology	Main epigenetic results	Phenotypic outcome	References
Oryzias melastigma (Medaka fish)	Dissolved oxygen	Hypoxia exposure of sexually mature (F0) fish	Global DNA methylation in sperm cells (F0–F2)	Methylated DNA immunoprecipitation with high-throughput sequencing (MeDIP-seq)	Global DNA hypermethylation in sperm of F2 fish; hypomethylation of promoter regions of *ehmt*2 and *ptk2b*; hypermethylation of exonic regions of *foxp*2	Transgenerational reproductive impairments such as aberrant sperm motility, and decreased testis weight, percentage of spermatids and percentage of fertilization success	Wang et al. (2016)
Three-spine stickleback *Gasterosteus aculeatus*	Salinity	Different environmental salinities	Genome-wide DNA methylation patterns between three-spines stickleback males and females and between stickleback reared at different environmental salinities	Whole genome bisulphite sequencing (WGBS)	Hypermethylation of X chromosome compared to Y chromosome		Metzger and Schulte (2018)
Culter alburnus	–	–	State of CpG methylation in the dmrt1 promoter of *C. alburnus* in the ovaries and testes	Bisulphite sequencing of the dmrt1 CpG island	Dmrt1 gene hypo-methylated and highly expressed in males	–	Jia et al. (2019)

Zebrafish (*Danio rerio*)	Elevated temperature	High temperature (33 °C) exposure of fish from 20 to 30 days post fertilization (dpf)	DNA methylation levels in *sox9b* and *esr1* promoter	Bisulphite sequencing and chromatin immunoprecipitation (ChIP) analysis	The promoters of sox9b and esr1 were hypermethylated by elevated temperature	High temperature-induced masculinization of zebrafish	Han et al. (2021)
Oyster *Crassostrea gigas*	NA	NA	Promoter methylation in the diacylglycerol kinase delta gene (*Dgkd*)	Bisulphite sequencing	Methylomes are related to infertility of triploid oysters	*Dgkd*: an epigenetic marker to distinguish male *C. gigas*	Sun et al. (2022)
Channel catfish *Ictalurus punctaus*	NA	Prior to sexual differentiation at 9, 12, and 16 dpf	NA	Bisulphite sequencing	Sex-specific epigenetically marked locus was identified in the sex determination region (SDR)	Epigenetic control of sex determination in channel catfish	Yang et al. (2022)
Ussuri catfish *Pseudobagrus ussuriensis*	NA	Sampling of female and male at the age of 2	NA	Methyl-RAD sequencing	Differential methylation patterns between males and females	*amh*, *gsdf* and *hsd11b2* in brain tissues and *slco3a1*, *socs2* and *trim47* in gonad tissues	Li et al. (2022)

influence muscle growth, gene expression and DNA methylation, demonstrating phenotypic plasticity (Campos et al. 2013). A study on young Atlantic cod (*Gadus morhua*) showed that increased growth due to changes in light exposure was linked to the altered activity of specific DNMTs in fast muscle tissue (Giannetto et al. 2013). These two studies suggested that DNA methylation could regulate the pace of muscle growth in response to temperature and photoperiod.

3.4.3 Sex Reversal

Some cases of developmental activities in aquaculture have been recorded, such as fish sex determination. High temperatures at an early stage in European sea bass were associated with a higher percentage of phenotypic males (Navarro-Martín et al. 2009). Navarro-Martin and colleagues found that exposure to high temperatures early in life was associated with increased DNA methylation in the promoter of the gonadal aromatase gene in adulthood (Navarro-Martín et al. 2011). In a separate work, faux males were induced by exposing genetic females to elevated temperatures during the sensitive embryonic phase of the half-smooth tongue sole (*Cynoglossus semilaevis*) (Chen et al. 2014). The DNA methylation profiles of pseudo males in the testes resemble those of genetic males, unlike the ovarian methylation patterns of healthy females. The fish species exhibits a pseudo-male phenotype and testis-specific methylation patterns inherited by the F1 generation, indicating transgenerational epigenetic inheritance of environmentally caused sex reversal events (Shao et al. 2014).

3.4.4 Epigenetic Inheritance

Due to environmental factors, brood-stock conditioning is a critical issue to consider in potentially transmitting epigenetic information from parents to offspring. Transgenerational plasticity and transgenerational epigenetic inheritance refer to the transfer of phenotypes to offspring through epigenetic processes due to environmental influences. Recent focus has been on training brood stock and transgenerational epigenetic inheritance in aquatic species like fish and shellfish, particularly concerning ecological fitness and climate change (Norouzitallab et al. 2014; Roy et al. 2019). Brine shrimp (*Artemia franciscana*) offspring in three consecutive generations (F1-F3) showed increased survival rates when faced with biotic stress (Vibrio challenge) and abiotic stress (thermotolerance) after their parental generation was subjected to heat shock or phloroglucinol, a plant-based compound that triggers heat shock, at early stages of development. A study conducted by Zhao et al. (2017) involved modifying the environment by exposing adult Manila clams and Sydney rock oysters to low pH conditions. This resulted in offspring that were faster-growing and better fit.

3.4.5 Epigenetics in Selection and Domestication

Epigenetic selection, independently or in combination with genetic selection, can be used in aquaculture to identify individuals with desired traits and enhance the chances of producing animals with desired characteristics (Gavery and Roberts 2017). Animal phenotypic changes can result from the interplay of the genome and the epigenome, indicating that genomic data may not thoroughly explain all variations in animal traits. Environmental epigenetic modifications can considerably impact individual variations and phenotypic outcomes across and between generations (Gavery and Roberts 2017; Roy et al. 2019). Epigenetic markers could be used in aquaculture to select brood stocks. A study on the Pacific oyster (*Crassostrea gigas*) investigated the genetic and epigenetic variability in mass selection populations used for selective breeding. The results suggested a potential link between the genetic environment of *C. gigas* and epigenetic variation (Jiang et al. 2013). Genome-wide methylation profiles were conducted to analyse the epigenetic differences between hatchery-raised Pacific salmon and wild salmon from two rivers. Epigenetic changes due to hatchery rearing may explain reduced fitness in hatchery-reared salmon released into the wild, as indicated by significant epigenetic variation in salmon (Le Luyer et al. 2017).

3.5 Conclusions

Our knowledge of epigenetics in aquatic organisms is currently limited, but there is potential for use in environmental modification to change desired traits for sustainable production. The epigenetics mark linked with production or performance traits could be utilized along with a genetic variation for the selective breeding of candidate aquaculture species. Although it is hypothesized that epigenetics has a significant impact on fish production, further studies are required to reveal their exact mechanism of gene regulation associated with vital traits. There is a need to better understand how epigenetic mechanisms and genotypes interact to influence the development of phenotypic features in response to environmental stressors. The recently available whole genome sequences and computational algorithms could facilitate mapping and detecting methyl marks in key genes and further could be useful in fish genomic research.

Acknowledgement The authors are thankful to Director, Indian Council of Agricultural Research - CIFA, Bhubaneswar, Odisha, India and Director, Indian Council of Agricultural Research - CIFE, Mumbai for facilitating this work.

Conflicts of Interest All the authors declared that no conflict of interest.

References

Abdelrahman H, ElHady M et al (2017) Aquaculture genomics, genetics and breeding in the United States: current status, challenges, and priorities for future research. BMC Genomics 18:191

Balasubramanian MN, Panserat S, Dupont-Nivet M, Quillet E, Montfort J, Le Cam A, Medale F, Kaushik SJ, Geurden I (2016) Molecular pathways associated with the nutritional programming of plant-based diet acceptance in rainbow trout following an early feeding exposure. BMC Genomics 17:449

Campos C, Valente LM, Conceição LE, Engrola S, Fernandes JM (2013) Temperature affects methylation of the myogenin putative promoter, its expression and muscle cellularity in Senegalese sole larvae. Epigenetics 8:389–397

Canário AVM (2019) Editorial: Aquaculture genomics. Aquacult Fish 4:1–2

Cavalli G, Heard E (2019) Advances in epigenetics link genetics to the environment and disease. Nature 571:489–499

Chen S, Zhang G, Shao C, Huang Q et al (2014) Whole-genome sequence of a flatfish provides insights into ZW sex chromosome evolution and adaptation to a benthic lifestyle. Nat Genet 46:253–260

Chu C-Y, Rana TM (2007) Small RNAs: regulators and guardians of the genome. J Cell Physiol 213:412–419

Cokus SJ, Feng S, Zhang X, Chen Z, Merriman B, Haudenschild CD, Pradhan S, Nelson SF, Pellegrini M, Jacobsen SE (2008) Shotgun bisulphite sequencing of the Arabidopsis genome reveals DNA methylation patterning. Nature 452:215–219

Dorts J, Falisse E, Schoofs E, Flamion E, Kestemont P, Silvestre F (2016) DNA methyltransferases and stress-related genes expression in zebrafish larvae after exposure to heat and copper during reprogramming of DNA methylation. Sci Rep 6:34254

Feng J, Chang H, Li E, Fan G (2005) Dynamic expression of de novo DNA methyltransferases Dnmt3a and Dnmt3b in the central nervous system. J Neurosci Res 79:734–746

Freitas MA, Sklenar AR, Parthun MR (2004) Application of mass spectrometry to the identification and quantification of histone post-translational modifications. J Cell Biochem 92:691–700

Gavery MR, Roberts SB (2017) Epigenetic considerations in aquaculture. PeerJ 5:e4147

Giannetto A, Nagasawa K, Fasulo S, Fernandes JM (2013) Influence of photoperiod on expression of DNA (cytosine-5) methyltransferases in Atlantic cod. Gene 519:222–230

González-Recio O (2011) Epigenetics: a new challenge in the post-genomic era of livestock. Front Genet 2:106

Green TJ, Helbig K, Speck P, Raftos DA (2016) Primed for success: Oyster parents treated with poly(I:C) produce offspring with enhanced protection against Ostreid herpesvirus type I infection. Mol Immunol 78:113–120

Han J, Hu Y, Qi Y, Yuan C, Naeem S, Huang D (2021) High temperature induced masculinization of zebrafish by down-regulation of sox9b and esr1 via DNA methylation. J Environ Sci 107:160–170

Harrison A, Parle-McDermott A (2011) DNA methylation: a timeline of methods and applications. Front Genet 2:74

Houwing S, Kamminga LM, Berezikov E, Cronembold D, Girard A, van den Elst H, Filippov DV, Blaser H, Raz E, Moens CB, Plasterk RH, Hannon GJ, Draper BW, Ketting RF (2007) A role for Piwi and piRNAs in germ cell maintenance and transposon silencing in zebrafish. Cell 129:69–82

Jablonka E, Raz G (2009) Transgenerational epigenetic inheritance: prevalence, mechanisms, and implications for the study of heredity and evolution. Q Rev Biol 84:131–176

Jaenisch R, Bird A (2003) Epigenetic regulation of gene expression: how the genome integrates intrinsic and environmental signals. Nat Genet 33(Suppl):245–254

Jia Y, Zheng J, Chi M, Liu S, Jiang W, Cheng S, Gu Z, Chen L (2019) Molecular identification of dmrt1 and its promoter CpG methylation in correlation with gene expression during gonad development in Culter alburnus. Fish Physiol Biochem 45:245–252

Jiang Q, Li Q, Yu H, Kong L-F (2013) Genetic and epigenetic variation in mass selection populations of Pacific oyster Crassostrea gigas. Genes Genomics 35:641–647

Jin B, Li Y, Robertson KD (2011) DNA methylation: superior or subordinate in the epigenetic hierarchy? Genes Cancer 2:607–617

Kelley JL, Tobler M, Beck D, Sadler-Riggleman I, Quackenbush CR, Arias Rodriguez L, Skinner MK (2021) Epigenetic inheritance of DNA methylation changes in fish living in hydrogen sulfide–rich springs. Proc Natl Acad Sci U S A 118:e2014929118

Khalaf O, Gräff J (2016) Structural, synaptic, and epigenetic dynamics of enduring memories. Neural Plast 2016:3425908

Le Luyer J, Laporte M, Beacham TD, Kaukinen KH, Withler RE, Leong JS, Rondeau EB, Koop BF, Bernatchez L (2017) Parallel epigenetic modifications induced by hatchery rearing in a Pacific salmon. Proc Natl Acad Sci U S A 114:12964–12969

Lee JT (2009) Lessons from X-chromosome inactivation: long ncRNA as guides and tethers to the epigenome. Genes Dev 23:1831–1842

Li R, Grimm SA, Wade PA (2021) CUT&Tag-BS for simultaneous profiling of histone modification and DNA methylation with high efficiency and low cost. Cell Rep Methods 1:100118

Li P, Chen J, Zhu C, Pan Z, Li Q, Wei H, Wang G, Cheng W, Fu B, Sun Y (2022) DNA methylation difference between female and male Ussuri Catfish (Pseudobagrus ussuriensis) in brain and gonad tissues. Life (Basel) 12:874

Liu Z, Zhou T, Gao D (2022) Genetic and epigenetic regulation of growth, reproduction, disease resistance and stress responses in aquaculture. Front Genet 13:994471

Lu G, Luo M (2020) Genomes of major fishes in world fisheries and aquaculture: status, application and perspective. Aquacult Fish 5:163–173

Metzger DCH, Schulte PM (2018) The DNA methylation landscape of stickleback reveals patterns of sex chromosome evolution and effects of environmental salinity. Genome Biol Evol 10:775–785

Moghadam HK, Johnsen H, Robinson N, Andersen Ø, Jørgensen EH, Johnsen HK, Bæhr VJ, Tveiten H (2017) Impacts of early life stress on the methylome and transcriptome of Atlantic salmon. Sci Rep 7:5023

Moore LD, Le T, Fan G (2013) DNA methylation and its basic function. Neuropsychopharmacology 38:23–38

Morán P, Marco-Rius F, Megías M, Covelo-Soto L, Pérez-Figueroa A (2013) Environmental induced methylation changes associated with seawater adaptation in brown trout. Aquaculture 392–395:77–83

Morgan HD, Santos F, Green K, Dean W, Reik W (2005) Epigenetic reprogramming in mammals. Hum Mol Genet 14 Spec No 1:R47–R58

Morris MJ, Monteggia LM (2014) Role of DNA methylation and the DNA methyltransferases in learning and memory. Dialogues Clin Neurosci 16:359–371

Narlikar GJ, Sundaramoorthy R, Owen-Hughes T (2013) Mechanisms and functions of ATP-dependent chromatin-remodeling enzymes. Cell 154:490–503

Navarro-Martín L, Blázquez M, Viñas J, Joly S, Piferrer F (2009) Balancing the effects of rearing at low temperature during early development on sex ratios, growth and maturation in the European sea bass (Dicentrarchus labrax): limitations and opportunities for the production of highly female-biased stocks. Aquaculture 296:347–358

Navarro-Martín L, Viñas J, Ribas L, Díaz N, Gutiérrez A, Di Croce L, Piferrer F (2011) DNA methylation of the gonadal aromatase (cyp19a) promoter is involved in temperature-dependent sex ratio shifts in the European Sea bass. PLoS Genet 7:e1002447

Ng K, Pullirsch D, Leeb M, Wutz A (2007) Xist and the order of silencing. EMBO Rep 8:34–39

Norouzitallab P, Baruah K, Vandegehuchte M, Van Stappen G, Catania F, Bussche JV, Vanhaecke L, Sorgeloos P, Bossier P (2014) Environmental heat stress induces epigenetic transgenerational inheritance of robustness in parthenogenetic Artemia model. FASEB J 28:3552–3563

Panserat S, Marandel L, Geurden I, Veron V, Dias K, Plagnes-Juan E, Pegourié G, Arbenoits E, Santigosa E, Weber G, Verlhac Trichet V (2017) Muscle catabolic capacities and global hepatic epigenome are modified in juvenile rainbow trout fed different vitamin levels at first feeding. Aquaculture 468:515–523

Pierron F, Baillon L, Sow M, Gotreau S, Gonzalez P (2014) Effect of low-dose cadmium exposure on DNA methylation in the endangered European eel. Environ Sci Technol 48:797–803

Rinn JL, Kertesz M, Wang JK, Squazzo SL, Xu X, Brugmann SA, Goodnough LH, Helms JA, Farnham PJ, Segal E, Chang HY (2007) Functional demarcation of active and silent chromatin domains in human HOX loci by noncoding RNAs. Cell 129:1311–1323

Roy S, Kumar V, Bossier P, Norouzitallab P, Vanrompay D (2019) Phloroglucinol treatment induces transgenerational epigenetic inherited resistance against vibrio infections and thermal stress in a brine shrimp (Artemia franciscana) model. Front Immunol 10:2745

Shao C, Li Q, Chen S, Zhang P, Lian J, Hu Q, Sun B, Jin L, Liu S, Wang Z, Zhao H, Jin Z, Liang Z, Li Y, Zheng Q, Zhang Y, Wang J, Zhang G (2014) Epigenetic modification and inheritance in sexual reversal of fish. Genome Res 24:604–615

Singh P, Szabó PE (2012) Chromatin immunoprecipitation to characterize the epigenetic profiles of imprinted domains. Methods Mol Biol 925:159–172

Sun D, Li Q, Yu H (2022) DNA methylation differences between male and female gonads of the oyster reveal the role of epigenetics in sex determination. Gene 820:146260

Uren Webster TM, Rodriguez-Barreto D, Martin SAM, Van Oosterhout C, Orozco-terWengel P, Cable J, Hamilton A, Garcia De Leaniz C, Consuegra S (2018) Contrasting effects of acute and chronic stress on the transcriptome, epigenome, and immune response of Atlantic salmon. Epigenetics 13:1191–1207

Wang SY, Lau K, Lai K-P, Zhang J-W, Tse AC-K, Li J-W, Tong Y, Chan T-F, Wong CK-C, Chiu JM-Y, Au DW-T, Wong AS-T, Kong RY-C, Wu RS-S (2016) Hypoxia causes transgenerational impairments in reproduction of fish. Nat Commun 7:12114

Yang Y, Zhou T, Liu Y, Tian C, Bao L, Wang W, Zhang Y, Liu S, Shi H, Tan S, Gao D, Dunham RA, Liu Z (2022) Identification of an epigenetically marked locus within the sex determination region of channel catfish. Int J Mol Sci 23(10):5471

Yuan Z, Liu S, Zhou T, Tian C, Bao L, Dunham R, Liu Z (2018) Comparative genome analysis of 52 fish species suggests differential associations of repetitive elements with their living aquatic environments. BMC Genomics 19:141

Zeng Y, Chen T (2019) DNA methylation reprogramming during mammalian development. Genes (Basel) 10:257

Zhao L, Schöne BR, Mertz-Kraus R, Yang F (2017) Sodium provides unique insights into transgenerational effects of ocean acidification on bivalve shell formation. Sci Total Environ 577:360–366

Proteomics in Fisheries and Aquaculture

4

Ajaya Kumar Rout and Bijay Kumar Behera

Abstract

Aquaculture is predicted to increase faster than other leading food production industries. It is also a central food-producing area contributing to food and nutritional security. Fish is essential to human nutrition, accounting for at least 20% of protein consumption for one-third of the world population. Aquaculture has encountered numerous challenges in achieving sustainable fish production due to its increasing importance. As a result, many omics technologies such as genomics, transcriptomics, and proteomics have been introduced to address various difficulties in fisheries. During the recent decade, there has also been an increase in the use of these Omics technologies in aquaculture. Advancements in proteomic techniques can help solve several concerns in aquaculture, such as growth, disease resistance, climate change, and aqua food safety and quality. This study evaluates basic and advanced proteomics technology and their potential use in tackling critical issues in aquaculture. Various proteomics approaches are explained, as well as existing bioinformatic resources and examples of their use in multiple areas of fisheries about safety, quality, and health. Proteomics can potentially be a valuable technique for analyzing the proteomes of any fish or food product. Through a comprehensive evaluation, this method enables the analysis of various traits of fish and the discovery of molecular indicators for their health, promoting sustainable fish growth and thus enhancing the quality and safety of aquatic food.

A. K. Rout (✉)
Rani Lakshmi Bai Central Agricultural University, Jhansi, Uttar Pradesh, India

B. K. Behera
College of Fisheries, Rani Lakshmi Bai Central Agricultural University, Jhansi, Uttar Pradesh, India

B. K. Behera (ed.), *Current Trends in Fisheries Biotechnology*,
https://doi.org/10.1007/978-981-97-3165-7_4

Keywords

Aquaculture · Proteomics · Animal protein

4.1 Introduction

Fish and fisheries products constitute the primary source of animal protein for the human diet, and aquaculture is the fastest-growing food sector. Foodstuffs are mostly made up of proteins, notably fish and meat, where proteins comprise most of the food matrix. Proteins regulate and support the formation of the food structure on the cell and tissue level. The growing population has led to a rise in the need for fish production. In 2018, the FAO documented 422 fish species being farmed, which included 7 finfish hybrids. Developing nations had the highest annual average individual fish consumption, reaching 20.5 kg (Kumar et al. 2019a, b, 2021a; FAO 2020). Fish and fishery products are a crucial part of the human diet and a great source of vitamins, vital and functional amino acids, ω-3-rich oils (particularly EPA and DHA), and high-quality animal proteins. Since the small native fish are rich in micronutrients, they can be crucial in the fight against disease caused by a lack of certain nutrients, common in developing and poor nations. For nearly 400 million individuals in emerging and impoverished countries, they provide more than half of their protein and mineral needs. Numerous nutraceutical and medicinal uses may be derived from fish oils, particularly their ω-3 PUFAs, EPA, and DHA content (Kumar 2022; Mohanty 2015; Sharma et al. 2019).

Proteomics is a potent tool in protein analysis that adds information on the location, ratio, and protein expression conditions to gene sequence data. Proteomic technologies have been employed more often in fish biology studies during the past 10 years. The main uses of proteomics have been in the study of fish model organisms like zebrafish (*Danio rerio*) and certain commercial species raised in aquaculture, mostly cyprinids and salmonids as well as their physiology, nutrition, health, quality, and food safety, as well as the impact of contaminants (Forne et al. 2010). Fish and fisheries products provide an average of 35 kcal per person every day worldwide, providing valuable nutritional resources with easily digestible animal proteins, which contributes to their widespread use (FAO 2020; Kumar et al. 2021b; Semwal et al. 2023).

Systems biology is a systematic method for comprehending biological processes at many levels, such as cell, tissue, and organism. Omics technology is used to study the entire genome, transcriptome, and proteomics. The suffix -ome originated in 1920 with the term genome, which describes an organism's complete genetic makeup (Yadav 2007; Singh et al. 2023a, b, c). Subsequently, a comprehensive study of genomics, transcriptomics, and proteomics was performed employing the suffix -omics (Volkova and Geras'kin 2018).

According to Schwanhäusser et al. (2011), the adaptable genome of an organism is essential as it offers insights into the complete gene expression system, encompassing post-transcriptional and post-translational expression of protein regulation.

Recent studies have demonstrated how proteomics is used in food science and technology. It is a useful tool for food scientists to gain valuable insights into how the quality of a product evolves during different processing and storage stages, as well as the composition of its basic materials. Recent developments in analytical omic research have allowed for the study of proteomes and peptidomes from many species, which has aided in the creation of accessible protein databases for protein identification, gene ontology, and phylogenetic comparisons using homology-driven methods (Piras et al. 2016).

Proteomics in aquaculture has become increasingly important in the past decade, helping it achieve its goal of expanding production while maintaining high-quality and sustainable results. Changes in protein composition at various life stages of fish species highlight the importance of developmental signaling pathways in response to environmental changes. Exploring fish proteomics can help analyze changes in the proteome due to factors including sickness, pollution, toxins, and temperature fluctuations that may disrupt the organism's metabolism (Tomanek 2011). The muscle is the most consumable part, and its development is affected by elements such as dietary habits, osmoregulation, immunity, and general metabolism. These systems impact the health of fish and thus influence the quality of meals prepared from fish. High-quality fish meat is produced through consistent physiological processes involving multiple cells and tissues. Examining proteomics at the molecular level can offer valuable insights into the transcription factors and signaling proteins that control fish growth patterns (Johnston et al. 2011; Jangpangi et al. 2023).

This review study emphasizes how proteomics contributes to advancing sustainable fish development and enhancing the quality and security of aqua food. The paper discusses the use of proteomics in fisheries and aquaculture, different proteomic methodologies, and existing bioinformatics resources.

4.2 Fundamentals Aquaculture Proteomic Techniques

Proteomics focuses on two critical areas: expression and functional proteomics. In addition to this proteome mining, post-translational modifications, structural proteomics, and protein-protein interaction are also widely studied under proteomics (Fig. 4.1). Most biological systems are controlled by a network of proteins whose interactions can determine certain biological activities (Monti et al. 2009). Functional proteomics is used when proteins interact to form temporary or lasting complexes to carry out certain functions. Proteins and their interaction partners can be separated from whole protein extracts using affinity-based techniques or immunoprecipitation based on ligand binding characteristics (Monti et al. 2009).

Functional and expression proteomics relies on fundamental proteomics methods such as protein separation and classification. Expression proteomics involves quantitatively evaluating the complete proteome or a specific group of proteins in samples with different expression levels. This method can reveal proteins related to a particular biological process or proteins that change their expression in response to harmful substances, diseases, or any stress. Proteomic methods i.e.,

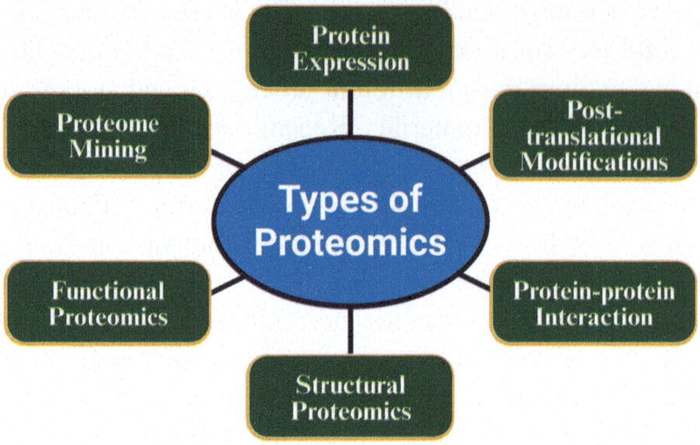

Fig. 4.1 Types of proteomics approach in fisheries and aquaculture

two-dimensional gel electrophoresis (2DE), protein microarray, chromatography, and mass spectrometry (MS) are employed for expression profiling (Vadivel 2015). A typical proteomic process involves sample preparation, protein digestion steps, protein separation methods, and identification algorithms. The choice of a particular stage in a proteomic analysis can differ based on the material type and the experiment's goal. This approach aims to isolate each protein in a biological sample to gather initial data for the later determination and quantification of proteins in future research. The proteomics process involves purification, detection, characterization, sequence analysis, quantification, and structural analysis, as shown in Fig. 4.2. Below are the approaches that can be used for proteome study in the fisheries or aquaculture industry.

4.2.1 Sample Preparation Concentrating on Fish Tissues

The preparation of samples is a crucial element of any proteomic research. The initial step in preparing a sample involves homogenization, which alters the sample's physical characteristics while maintaining its chemical properties. Cells need to be efficiently disrupted to eliminate all intracellular proteins. Various procedures are routinely used for preparing proteomic samples, including ultrasonic, osmotic, pressure-based, mechanical, freeze-thawing, detergent lysis, and others. The type of disruption chosen depends on the source of the sample (solid tissue, cell suspension, and other biological material) and whether the investigation targets all proteins or a specific subcellular subset (Bodzon-Kulakowska et al. 2007; Singh et al. 2023b). Sonication can disrupt tissue samples, i.e., kidney, muscle, liver, and spleen. For hard tissues like fin, skin, and scale, increased mechanical force, treatment time, and a particular lysis buffer are necessary for optimal disruption or homogenization (Kessels et al. 2014; Singh et al. 2023c). A low-power ultrasound can be used for cell lines and liquid biological specimens like serum, plasma, and mucus. Osmotic or detergent lysis efficiently disrupts mitochondrial and nuclear membranes in

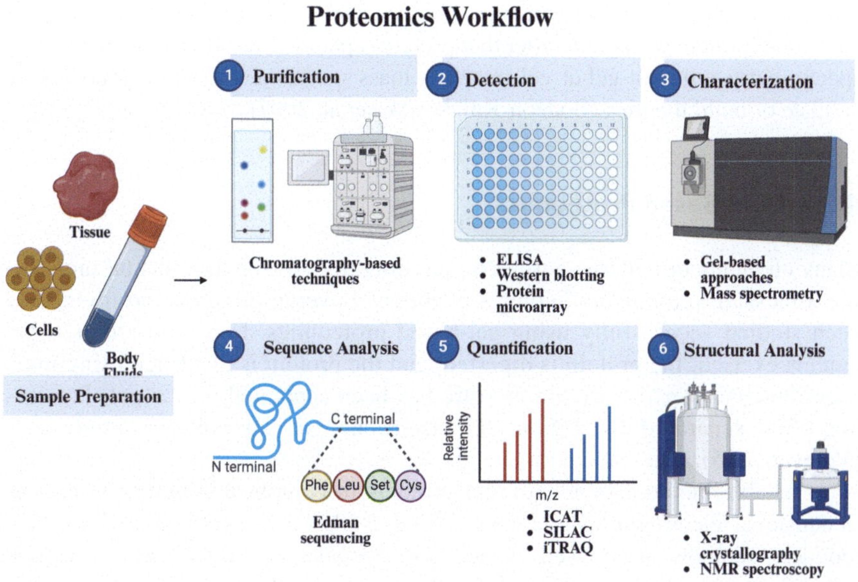

Fig. 4.2 An outline of the workflow followed in proteomics analysis. The proteomics workflow can be summarized into purification, detection, characterization, sequence analysis, quantification, and structural analysis

cellular extract. Typically, a protease inhibitor cocktail (PIC) is utilized during sample preparation to block protein degradation by endogenous proteases (Liu et al. 2010).

Proteins are initially insoluble and must destroy their connections to become soluble once separated. Detergents such as CHAPS and Triton X, chaotropic agents like protease inhibitors, thiourea and urea, and reducing agents can be utilized in solubilization. The solubilization buffer can be customized to suit the requirements of the experiment. An ionic solution is capable of extracting proteins from fish muscle. A high-ionic strength solution extracts muscle myofibrillar proteins, while a low-ionic strength solution extracts sarcoplasmic proteins (Borah et al. 2023). Furthermore, in the case of functional proteomics, when the primary goal is to characterize protein-protein interactions and investigate natural protein complexes, detergent selection becomes even more critical. According to Reisinger and Eichacker (2008), non-ionic mild detergents, i.e., dodecyl-maltoside, triton-x, and digitonin are commonly used to solubilize membrane proteins. The absence of contaminants in the sample is critical for proteomic sample preparation. Various procedures, including spin dialysis, affinity chromatography, ultracentrifugation, precipitation, and gel filtration chromatography can remove contaminants depending on their nature.

Various factors such as cell disruption method, protein content, solubilization methodology, and detergent choice impact the quality of sample preparation. Furthermore, depending on whether the downstream processing is gel-based or gel-free, the choice of reagents and buffers may change since various buffers work better in each situation. As of yet, there isn't an extraction buffer that works with all types of samples. The usual proteomic study workflow involves sample homogenization and protein extraction, after that protein separation and digestion using mass spectrometry, either in-gel or gel-free. The mass spectrometric data is evaluated to validate or build theories (Bodzon-Kulakowska et al. 2007).

4.2.2 Gel-Based Proteomics

Many elements of fisheries research including toxicity, environmental monitoring, illness, developmental biology, health, and overall proteome profiling have been studied successfully using gel-based proteomics. Here, the required gel band is excised, the protein is digested, and the protein is identified using mass spectrometry after the protein mixture has been separated on the gel. Proteins are often separated by 1D or 2D electrophoresis on polyacrylamide gels (Vadivel 2015).

A reliable method for identifying proteins in complicated materials is one-dimensional electrophoresis (1-DE). Using 1-DE (10% acrylamide) liver proteomics, rainbow trout were tested for *Aeromonas salmonicida* infection. Label-free quantification allowed for the detection of over 3000 proteins. To determine how an EDTA-supplemented diet affected allergenicity, 1D electrophoresis was used to analyze the muscle proteome of seabream. The most popular gel-based technique for protein expression and identification study of hundreds of proteins on a single gel is called 2D gel electrophoresis. Proteins are categorized according to their molecular weight in the second dimension and their isoelectric point in the first (Meleady 2018). The obtained gels are further stained to compare and see the protein spots and gel profile under different experimental settings. Bioinformatic software is used for spot identification, quantification, and is identical with other groups of gels (Abdallah et al. 2012). The proteome changes in Pacu twitch muscle during the fasting-refeeding diet were identified using the 2D electrophoresis technique. It has been used to study several other aspects related to fish, including authentication, identifying protein biomarkers for different infections or diseases, the effects of environmental factors, and the toxicological effects of pollutants. Furthermore, the widespread use of 2D electrophoresis in fisheries attests to its continued status as a critical method in fundamental science (Xu et al. 2019).

Aside from its many benefits, 2D has several disadvantages, including variances across users and gels, distortions in the gel, and problems with spot normalization. A more sensitive and precise approach for quantitative analysis called 2D-difference in gel electrophoresis was created as an extension to conventional 2D electrophoresis to get around these problems. In DIGE, labeling is done before the first-dimensional phase using three spectrally resolvable fluorescent tags (Meleady 2018). A novel diet formulation and the influence of varying diets on fish allergenicity and muscle quality have been studied using 2D-DIGE (Richard et al. 2016).

4.2.3 Proteomics Performed Without the Use of Gels

Gel-based proteomics has various drawbacks, including the inability to see the whole proteome on gel due to technical variances or further physicochemical features such as isoelectric point, hydrophobicity, and molecular weight. A standard 2DE gel can detect around 30–40% of the whole proteome (Abdallah et al. 2012). Gel-free proteomics is beneficial because it involves separating proteins and digestion without gel. HPLC is a leading proteomic separation method that may be seamlessly used with mass spectrometry to separate peptides or proteins before mass analysis. HPLC is the last step in the proteome analysis that can be used in either a top-down or bottom-up strategy.

In the top-down technique, protein mixtures are divided and specific proteins or protein mixes are selected for mass spectrometry. Peptides derived from protein digests are analyzed using mass spectrometry in the bottom-up proteomic method (Vorontsov et al. 2016). Peptides are fractionated in this investigation, resulting in the discovery of several peptides and subsequently proteins. Liquid chromatography is employed to separate the peptide mixture prior to analysis with mass spectrometry, rather than utilizing polyacrylamide gels. Shotgun proteomics allows for the quantification and recognition of proteins by studying peptides generated from protein degradation. Shotgun proteomics has improved the total protein identification in fish (Piovesana et al. 2016).

Gel-free techniques are faster and need smaller sample sizes compared to gel-based procedures. While the gel-free technique offers notable benefits, it is not a substitute for the gel-based approach. Both methods are considered supplementary, and the research issue may influence the selection of a specific methodology.

4.2.4 Protein Digestion Approaches

Protein digestion is crucial for accurately identifying proteins, regardless of whether the method is gel-free or gel-based. Protein digestion can occur within a gel matrix, in a liquid solution, or on a filter. In-gel digestion isolates specific gel bands or patches and eliminates the stain. The proteins in the band/spot undergo reduction, digestion, and alkylation within the gel. Peptides are ultimately separated and examined by LC-MS/MS. To investigate the development process of transgenic and selective breeding fish, in-gel digestion was employed to detect and characterize the proteome differences (Causey et al. 2019). All stages of protein digestion in a liquid solution, such as denaturation, alkylation, reduction, and digestion, occur in the liquid phase. The protein samples are purified to remove salts before being analyzed using mass spectrometry. This digesting method is becoming increasingly popular and is often used for preparing samples. The muscle proteins of wild and farmed gilthead sea bream were analyzed by in-solution digestion (Piovesana et al. 2016).

The FASP method is utilized for on-filter digestion to eliminate interfering compounds of both high and low molecular weights. The necessary procedures, such as digestion, are carried out on the filter, resulting in the eventual retrieval of the protein digests. White perch membrane and cytosolic proteins were evaluated using an on-filter digestion technique. The goal was to identify proteins that were expressed differently in rainbow trout lymphoid organs after a bacterial infection (Kumar et al. 2018).

4.2.5 Protein Identification Using Mass Spectrometry Data Collection

Mass spectrometry (MS) is generally the penultimate stage in a proteomic investigation, providing data for protein identification and quantification. Proteomic analysis uses mass spectrometry as a primary approach. Urban (2016) describes the principle on which a mass spectrometer operates. Ions are produced from the analyte sample, separated in a mass analyzer according to their mass-to-charge (*m*/*z*) ratio, and the proportion of each ion type is recorded. A mass spectrometer typically consists of an ionization source, mass analyzer, and detector. The substance leaving the high-performance liquid chromatography (HPLC) system becomes ionized and is introduced into the mass spectrometer in an HPLC-tandem mass spectrometry (HPLC-MS/MS) configuration to acquire raw mass spectrometry data. Two main categories of acquisition methods are data-independent (DIA) and data-dependent (DDA). DDA quantification is based on either spectrum counts or precursor ion intensity. The survey scan (MS1) selects the most intense precursors for tandem fragmentation (MS2). Data-independent acquisition (DIA) does not rely on the prior detection of precursor ions or any information about the mass-to-charge (*m*/*z*) values of precursors to acquire fragment ion spectra (MS2), unlike other approaches (Ludwig et al. 2018).

By integrating DIA with traditional DDA-MS, identifying proteins has been successfully enhanced, improving sensitivity, accuracy, and reproducibility. Initially, conventional DDA-MS was used to investigate and develop a thorough proteome profile or construct a library. The DDA technique was utilized for protein identification and quantification in food fish: proteome profiling of common carp during bacterial infection and creating gill assay libraries in three spine stickleback species (Monroe et al. 2020).

Before entering the MS, analytes must be ionized to be detected. The two main soft ionization methods for proteomics applications are matrix-aided laser desorption ionization (MALDI) and electrospray ionization (ESI), which ionize peptides by adding or removing protons. According to Singhal et al. (2015), MALDI predominantly generates monovalent ions of the sample that are subsequently separated according to the *m*/*z* ratio. Using electrical energy, ions are transferred from the solution phase to the gas phase in the case of ESI, allowing for the separation and detection of numerous charged ions. Because ESI may produce multiplex charged ions, it can effectively increase the mass range of analyzer to accept proteins of varying sizes and the polypeptide fragments that go along with them. Depending on the type of ionization and mix of analyzers it will be employing, a mass spectrometer can be set up in various ways. Mass spectrometric analysis offers insights into identifying proteins and quantification, which can be broad or specific depending on the data collection method, instrument setup, and data processing technique.

4.2.6 Quantitative Proteomic Approaches

Depending on the proteomic separation method, gel-based or non-gel-based quantitative techniques are available. Two types of gel-based quantification approaches include conventional 2DE and sophisticated 2D-DIGE. They may also be divided into two categories: label-based quantification techniques and label-free quantification (LFQ). 2DE falls under LFQ among gel-based techniques, whereas DIGE is a label-based proteomic method that tags proteins with cyanine dyes. Label-based techniques lower the predicted variability if individual samples are handled by labeling samples, pooling them, and then exposing them to MS. Here, the stability of the isotope labels utilized determines the protein quantity (Latosinska et al. 2015; Mavi et al. 2023).

Label-based and label-free methods have been used in many studies on fish quality, health, production, and metabolism. The TMT approach was used to detect changes in the proteome of grouper (*Epinephelus fuscoguttatus*) fillets as they were stored in a refrigerator. This helped in pinpointing potential protein markers for meat quality. An iTRAQ-based quantitative analysis in grass carp showed that consuming a faba bean diet can induce apoptosis and hemolysis. LFQ-based mass spectrometry was utilized to analyze changes in the liver proteome of rainbow trout following infection with *Aeromonas salmonicida*. The LFQ method was used to identify and characterize putative protein indicators for tenderness in Tilapia fillets (He et al. 2018).

4.2.7 Specific Proteomic Methods

Targeted proteomics is a method used as an alternative to immunoassay for the specific identification of proteins and peptides. This advanced proteomic approach is used in system biology studies across many omics (Manes and Nita-Lazar 2018). Specific fish or food proteins can be accurately and repeatedly identified in a customized proteomic study. The predominant targeted method is chosen reaction monitoring (SRM), sometimes known as multiple reaction monitoring (MRM). A QQQ mass spectrometer is used for SRM. In this scenario, the instrument approach is set up to provide mass-to-charge ratio data for both the precursor and relevant fragment ions. Each peptide can exhibit many transitions in a single SRM experiment. A transition is a combination of a precursor and a fragment ion. This approach is unique because the likelihood of fragment ions from two isobaric peptides having the same retention time is exceedingly low. SRM is a productive technique that effectively finds protein peptides in biological samples, even those with low abundance (Borràs and Sabidó 2017).

According to Shi et al. (2016), two further somewhat sophisticated targeted methods are sequential windowed acquisition of all theoretical fragment ion mass spectra (SWATH-MS) and parallel reaction monitoring (PRM). As a result, the fragment ions are identified with higher mass accuracy and resolution than SRM because these are often carried out using Q-TOF or Q-OT instrument setups. In PRM, if the instrument configuration is Q-OT, the first mass analyzer, or quadrupole, separates the predetermined peptide and sends it via C-trap to the collision cell (HCD, or high energy collision dissociation cell), where it is fragmented and all of the fragmented ions are sent to the detector. In PRM, peptide selection is carried out and product ion selection is unnecessary, which sets it apart from SRM. The parallel reaction monitoring has good specificity and sensitivity for protein measurement (Borràs and Sabidó 2017).

SWATH-MS employs the data-independent acquisition (DIA) method to achieve comprehensive proteome analysis with high precision and reliability, integrating advantages from shotgun and selected reaction monitoring (SRM) approaches. For precursor ion selection in SWATH-MS, an acquisition window is established to encompass the mass range where most peptide precursors in a biological sample are expected to be found. The precursor ions are sequentially selected inside the isolation window and subjected to collision activation, and all resulting product ions from all precursors are directed to the detector (Borràs and Sabidó 2017). SWATH-MS data provide a comprehensive and efficient map of all fragment ions within a specific mass range. A peptide-centric scoring technique is required to assess this data, along with a spectrum library and prior knowledge of peptide chromatographic and mass spectrometric behavior (Ludwig et al. 2018). SRM, PRM, and SWATH workflows have been standardized (Sakshi et al. 2023). Targeted proteomic approaches can validate specific protein indicators for species identification, food quality, or sickness. An SRM-based approach was used to identify rainbow trout adulteration in Atlantic salmon to authenticate processed food products and address batch variability. A PRM-based method was employed to monitor parasite-specific peptide markers in fish samples to detect parasite anisakids (Carrera et al. 2016).

SWATH-MS is an innovative method for gathering high-throughput proteome data across different species. Analyzed SWATHMS data from rainbow trout stomach and lymphoid tissue after Yersinia ruckeri infection using spectrum libraries derived from matched DDA-MS data. The impact of high-pressure processing on muscle quality and efficiency in European sea bass fillets was evaluated using SWATH-MS. Similar targeted proteomic methods can generate large-scale data to identify potential quality or authenticity protein markers in fish or other food items for developing and establishing validation protocols (Kumar et al. 2018, 2019a, b).

4.3 Bioinformatics Resources to Improve Proteomic Data Analysis

Computational and bioinformatic tools are essential for fully implementing proteomics methods in any discipline. For instance, bioinformatics is used throughout the whole proteomic experiment, from raw data collection to processing and interpretation using mass spectrometry. These platforms connect extensive omics or multi-omics data to study the biological processes within a living organism (Mohanty et al. 2019). Automatic search engines like Sequest, Mascot, Comet, and MaxQuant (Andromeda) are crucial for identifying and quantifying peptides and proteins in raw MS data. Protein sequences from genomic or transcriptomic data are frequently present in these databases for identifying proteins. NCBI possesses genomic sequence data for 331 fish species and 24 species used in aquaculture (Yue and Wang 2017).

Pathway and network analysis is an advanced method that enhances the comprehension of scientific information to reach a meaningful conclusion. The pathway library includes data on proteins or genes that are involved in the pathway, as well as those with a regulatory influence or participating in reactions inside the pathway. Various pathway and network evaluation tools are available, including DAVID, KEGG, and IPA. Reactome, g:Profiler, and GSEA (Cirillo et al. 2017). Reactome and g:Profiler provide databases for Zebrafish, while KEGG includes a variety of food fishes and a small number of other fish species.

Several proteins require stable or transitory compounds to function correctly. The process of synthesis or dissociation of these complexes should be investigated to understand their biological activities better. All of this is connected to interactions between proteins and various interaction databases provide experimentally determined or anticipated information regarding protein interactions. STRING and Cytoscape are two often used tools for protein-protein interactions. STRING includes interaction information and links to a variety of different educational materials. Different experimental datasets and interaction networks may be constructed in Cytoscape (Doncheva et al. 2019). These methodologies are components of systems biology, which aids in understanding biological systems.

4.4 Proteomics Applications in Fisheries and Aquaculture

Aquaculture is expanding faster than any other food-producing industry on the planet. As a result, we must guarantee that farm-raised fish meet the same standards as wild fish. This may appear to be an unexpected component to investigate initially, yet various factors can influence a fish's physiology. The increasing significance of food fish in ensuring nutritional security and nutrition has redirected attention from experimental fish species to food fishes and farmed species. Proteomic research has been used to study various aquaculture problems and fundamental biology through model fish species (Kumar et al. 2022a, b).

Proteomics has long been criticized as a time-consuming and costly technique, however, advances in protocols have resulted in faster and more reproducible findings. Additionally, the lower cost of proteomic technology has enabled its widespread use in academics and industry. Proteomics has been utilized to investigate various concerns in fisheries, including welfare, safety, quality, health, and diseases. Proteomic technology can aid in appropriately verifying food items and ensuring customer safety and quality (Fu et al. 2018).

4.4.1 Traceability and Authentication

Traceability and originality are essential for food scientists and the food sector to maintain public health and market transparency. Accurate species and origin identification of food items can help prevent fraud and illegal trade (Power and Cozzolino 2020). Microbe-food contact plays a crucial role in defining food product traceability. Molecular techniques aid in differentiating, detecting, and identifying microbes connected with food (El Sheikha et al. 2018).

Applications of several analytical methods for aquaculture traceability were evaluated. These biological and physicochemical tests examined certain components (antioxidants, fatty acids, and microbial DNA) in food items under the theory that these markers indicated a product's unique place of origin. The significance of the PCR-DGGE technology in generating a distinct barcode for every food item was highlighted (El Sheikha and Xu 2017).

In particular, whether working with fillets or minced fish, the proteomics method offers a viable substitute for safety, authenticity, and traceability. The Society of Certified Analytical Chemistry acknowledges isoelectric focusing analysis of sarcoplasmic proteins as a validated method for species identification. This technique has been used to distinguish between several fish and shrimp species. Six commercial species of shrimp that are regionally distinct were identified by creating a mass spectra library through MALDI-MS-based proteomics (Ortea et al. 2016).

In contrast to the current practice in many nations of employing whole animals or plants, cellular agriculture is a young area that uses cell culture to produce various agricultural products. In vitro models of toxicity, pathology, physiology, and gene expression levels are also studied using fish cell lines. Proteomics might be helpful in identifying a particular cell line by analyzing protein expression signatures (PES), in addition to DNA fingerprinting and PCR-based techniques. A protein unique to a particular state and can help examine and recognize variations across different cell or tissue types is referred to as PES. Proteomics techniques may therefore be helpful in accurately identifying fisheries and food items, preventing mislabeling, and guaranteeing the product's quality (Chen et al. 2019).

4.4.2 Evaluating Quality and Identifying Allergies

The quality of fish is essential because it indicates the changes that occur as an outcome of handling, nutritional supplements, and either pre- or post-mortem processing. The quality of fish is essential because it shows the changes that occur as a result of dietary supplements, conduct, and pre- or post-mortem storage and might be used as freshness indicators. Anisakis simplex, a parasite found in fish, is to blame for human anisakiasis and allergic responses worldwide. Proteome analysis of Anisakis larval stages L3 and L4 was conducted using TMT-based molecular quantification, resulting in eight protein profiles for L3 and 12 for L4 (Stryiński et al. 2019).

4.4.3 Nutritional Control

According to integrated omics and chemical investigations, there are notable variations between the fish meat from farmed and unfarmed sources, which variations in diet components may cause. Diets with anti-nutritional elements can harm fish growth and physiology by disrupting digestion and intestinal function. The fish feed formulation should be done carefully so that the supplements added do not harm the fish. Fish reaction to dietary supplements and other external stimuli has been studied using proteomics; the results of this research may be taken into account while formulating and optimizing fish feed. Thus, the impact of a specific diet on fish physiology may be comprehended by applying proteomics methods. Optimizing feed selection based on suitability with metabolism in fish is essential for maintaining a strong relationship between feed conversion efficiency and product quality (Krogdahl et al. 2015).

4.4.4 Health Management

A crucial component of aquaculture welfare is fish health, which is impacted by any negative environmental change, including stress, impacting human health. Enhancing fish health is the first step in producing healthy fish, which ensures the end product's quality and consequently, the health of the customer. Proteome profile characterization can aid in the study of immunological response and pathophysiology; as a result, protein markers can be adjusted to assess fish health. In this regard, proteome investigations concentrate specifically on skeletal anomalies, bacterial and viral infections, parasite invasions, and vaccine development.

Proteomics serves as a vital adjunct to traditional methods such as identifying pathogens, examining disease symptoms, and conducting histopathological evaluations for effective disease diagnosis in aquaculture. Thus, proteomics helps in the identification of pathogens, characterization of toxins, and studying host-pathogen interaction (Fig. 4.3).

Indicators of parasite infection (*Ichthyophthirius multifiliis*) in common carp skin mucus included label-free proteomics, which showed the downregulation of cytoskeletal proteins and over-expression of several metabolic proteins. These findings may have resulted from poor healing during the parasite infection (Saleh et al. 2018). When *Streptococcus agalactiae* infects the spleen of tilapia, the pathogen uses the toll receptor pathway to evade the host immune system, while the host uses NOD receptor signaling to combat the infection. This was discovered using proteomics in conjunction with other omics strategies (Wu et al. 2019). Rainbow trout infected with *A. salmonicida* showed elevated levels of acute phase response and complement system proteins in their livers, as demonstrated by label-free proteomic profiling.

Certain fishes are also prone to skeletal abnormalities. Apolipoproteins, calmodulin, MYL, and CKM were among the proteins linked to skeletal abnormalities in rainbow trout larvae (Babaheydari et al. 2016). Therefore, the use of proteomic

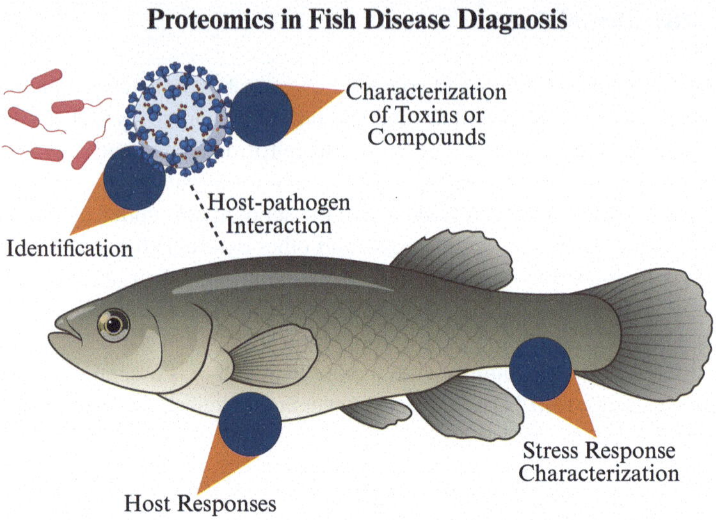

Fig. 4.3 Role of proteomics in fish disease diagnosis and understanding host-pathogen interaction

techniques may aid in a deeper comprehension of a disease's molecular process, which may aid in the development of disease-controlling agents or enhance fish farming practices.

4.4.5 Reproductive Study

Studies about issues of reproduction, fish reproduction, and development-related molecular pathways have been investigated by the application of proteomics. Fish proteomics has mostly concentrated on gametogenesis and the creation of innovative biomedical uses. Researching this topic will also benefit aquaculture and fisheries generally since high-quality seeds produce healthy individuals and a thorough understanding of the molecular underpinnings of reproduction will aid in the expansion of aquaculture. The quality of an animal's eggs impacts its total reproductive fitness (Yilmaz et al. 2017). ApoA1, Vitellogenin, and MBL proteins of oocyte quality are examples of possible protein indicators that were found in rainbow trout ovarian fluid using shotgun proteomics (Nynca et al. 2015). According to iTRAQ-based research, IDS proteins and HSP-70 may be useful indicators for assessing the quality of eggs in captive hapuku (*Polyprion oxygeneios*) (Kohn et al. 2015).

Furthermore, caviar and fish roe, or eggs are wholesome sources of fats and proteins. While roe may be found in the eggs of trout, carp, salmon, and a few other species. Caviar is the costliest meal made from sturgeon eggs (Vilgis 2020). Fish gametes are studied at the molecular level to learn more about their reproductive biology and find quality indicators for these meals. Female sturgeon egg proteome was analyzed using 2DE and MALDI-TOF-TOF. This data can serve as a starting point for investigating the molecular indicators for caviar quality (Keyvanshokooh and Vaziri 2008). These proteomic platform-based discoveries can add to our understanding of fish reproduction success as well as the preservation of the royalty and caliber of these delicacies.

4.4.6 Environment and Climate Change

Fish are subject to several stimuli that impact their development and general health, much like higher vertebrates. Fish handling techniques, overcrowding, osmotic conditions, pH, oxygen and carbon dioxide levels, and temperature are just a few of the variables that might harm fish due to climate change. Aquaculture welfare standards might be raised by monitoring stress and optimizing the fish stress response. Any of these stressors can impact metabolism and protein expression patterns, which can be reflected at the molecular level. Using proteomics, the degree of degradation brought on by a situation may be determined, and appropriate recommendations could then be developed to either lessen or eliminate the effects of stress. A study was carried out to evaluate protein-level changes in gilthead seabream exposed to three stressors: net handling, hypoxia, and overcrowding. The results showed that unique groups of proteins had varying levels of expression in each scenario (Raposo de Magalhães et al. 2020; Sharma et al. 2022). Proteomics can aid in comprehending the molecular mechanisms and physiological responses of fish to stress. It can also be used as a screening method to identify possible stress indicators, which might lead to the manufacture of stress-free fish.

4.4.7 Aquatic Environmental Toxicology

It is crucial to develop quick and accurate methods for detecting toxicity because various industrial wastes, such as toxic sludge, hazardous solvents, and heavy metals, are discharged into aquatic ecosystems annually. These releases affect aquatic organisms, which disrupts the entire food chain. The advancement of several omics platforms presents a potential means of monitoring the aquatic ecosystem (Mateo-Sagasta et al. 2017).

Quantitative proteomics of rainbow trout liver in their early life stages showed that exposure to the herbicide Diquat dibromide greatly enhanced caspase activity and activated the Atk/mTOR signaling pathway (McCuaig et al. 2020). The proteome changes in *Oryzias mellastigma* in response to ammonia stress were investigated using gel-based proteomics in conjunction with mass spectrometry (Zhu et al. 2018). The liver was subjected to proteomic analysis after treating grass carp (*Ctenopharyngodon idellus*) with the antibiotic enrofloxacin. The results indicated that the treatment impacted some biological processes, including focal adhesion, protein digestion, absorption, and interactions between the extracellular matrix (ECM) and receptors.

Proteomics may be used in this industry to examine the impacts of harmful substances on aquaculture health, screen for risk factors in consumers, and develop safety protocols to minimize negative consequences.

4.5 Conclusions

The increasing necessity to create and implement efficient methods for assessing the health of fish and quality arises from the increasing demand for more secure and nutritious food products. The food needs of an expanding population must be met through sustainable aquaculture practices. Fish welfare molecular indicators may be found by analyzing many features of fish using omics platforms, including transcriptomics, proteomics, and genomes. Here, we made an effort to evaluate and critically examine how proteomics has been applied as a potent tool for tackling a range of aquaculture-related challenges, including nutrition, quality, safety, and disease prevention. Fish proteomics research is still in its infancy, and there are several challenges to be addressed, such as the scarcity of certain reagents and the paucity of transcriptome and genomic data. Furthermore, information on a limited number of fish species has been integrated into some online databases. Studies on various edible fish species models have demonstrated the promise of proteomics in comprehending fish biology, tackling aquaculture obstacles, and identifying possible indicators for safety and quality. Identification and targeted proteomics provide precise methods for identifying the genuineness, tracking, and allergen identification in fish or fishery products.

Proteomic techniques can be applied to fish or any other dietary item, as proteins are considered the main functional biomolecules in all living organisms. In this article, we've highlighted some proteomic sample preparation protocols, ranging from simple to complex, along with the related methods that may be useful when researching how aquaculture and its surroundings affect food quality and safety in fishing products. Given that fish and food proteomics have the potential to develop into effective analytical tools for the food sector, this study highlights the necessity for additional ongoing research in this area. Furthermore, the blue revolution will be brought about in large part by the integration of proteomics data and other omics data.

References

Abdallah C, Dumas-Gaudot E, Renaut J, Sergeant K (2012) Gel-based and gel-free quantitative proteomics approaches at a glance. Int J Plant Genomics 2012:494572

Babaheydari SB, Keyvanshokooh S, Dorafshan S, Johari SA (2016) Proteomic analysis of skeletal deformity in diploid and triploid rainbow trout (*Oncorhynchus mykiss*) larvae. Comp Biochem Physiol Part D Genomics Proteomics 19:1–7

Bodzon-Kulakowska A, Bierczynska-Krzysik A, Dylag T, Drabik A, Suder P, Noga M, Jarzebinska J, Silberring J (2007) Methods for samples preparation in proteomic research. J Chromatogr B Anal Technol Biomed Life Sci 849:1–31

Borah D, Goswami RK, Shrivastav AK, Giri P, Kumar N, Singh T (2023) Recent trends of nano-biosensors in agriculture. In: Chaturvedi VK, Yadava DP, Singh MP (eds) Recent advances in biosensor technology, vol 2(14). Bentham Science, Sharjah, pp 152–165

Borràs E, Sabidó E (2017) What is targeted proteomics? A concise revision of targeted acquisition and targeted data analysis in mass spectrometry. Proteomics 17:17–18

Carrera M, Gallardo JM, Pascual S, González ÁF, Medina I (2016) Protein biomarker discovery and fast monitoring for the identification and detection of Anisakids by parallel reaction monitoring (PRM) mass spectrometry. J Proteome 142:130–137

Causey DR, Kim JH, Stead DA, Martin SA, Devlin RH, Macqueen DJ (2019) Proteomic comparison of selective breeding and growth hormone 35 transgenesis in fish: unique pathways to enhanced growth. J Proteome 192:114–124

Chen B, Zheng Z, Yang J, Chi H, Huang H, Gong H (2019) Development and characterization of a new cell line derived from European eel *Anguilla Anguilla* kidney. Biol Open 8:bio037507

Cirillo E, Parnell LD, Evelo CT (2017) A review of pathway-based analysis tools that visualize genetic variants. Front Genet 8:174

Doncheva NT, Morris JH, Gorodkin J, Jensen LJ (2019) Cytoscape StringApp: network analysis and visualization of proteomics data. J Proteome Res 18:623–632

El Sheikha AF, Xu J (2017) Traceability as a key of seafood safety: reassessment and possible applications. Rev Fish Sci Aquacult 25:158–170

El Sheikha AF, Levin RE, Xu J (2018) Molecular techniques in food biology: safety, biotechnology, authenticity and traceability. Wiley, New York

FAO (2020) The state of world fisheries and aquaculture 2020. In: Sustainability in action. FAO, Rome

Forne I, Abia NJ, Cerda J (2010) Fish proteome analysis: model organisms and non-sequenced species. Proteomics 10:858–872

Fu Q, Kowalski MP, Mastali M, Parker SJ, Sobhani K, Van Den Broek I, Hunter CL, Van Eyk JE (2018) Highly reproducible automated proteomics sample preparation workflow for quantitative mass spectrometry. J Proteome Res 17:420–428

He Y, Huang H, Li LH, Yang X (2018) Label-free proteomics of tilapia fillets and their relationship with meat texture during post-mortem storage. Food Anal Methods 11:3023–3033

Jangpangi K, Khati A, Chauhan RS, Kumar N (2023) Assessment of the haematological and biochemical effects of Himalayan herb *Urtica dioica* leaves diets fed Amur carp, *Cyprinus carpio haematopterus*. Turk J Fish Aquat Sci 23(12):TRJFAS23854

Johnston IA, Bower NI, Macqueen DJ (2011) Growth and the regulation of myotomal muscle mass in teleost fish. J Exp Biol 214:1617–1628

Kessels MY, Huitema LF, Boeren S, Kranenbarg S, Schulte-Merker S, van Leeuwen JL, de Vries SC (2014) Proteomics analysis of the zebrafish skeletal extracellular matrix. PLoS One 9:e90568

Keyvanshokooh S, Vaziri B (2008) Proteome analysis of Persian sturgeon (*Acipenser persicus*) ova. Anim Reprod Sci 109:287–297

Kohn YY, Symonds JE, Kleffmann T, Nakagawa S, Lagisz M, Lokman PM (2015) Proteomic analysis of early-stage embryos: implications for egg quality in hapuku (*Polyprion oxygeneios*). Fish Physiol Biochem 41:1403–1417

Krogdahl Å, Gajardo K, Kortner TM, Penn M, Gu M, Berge GM, Bakke AM (2015) Soya saponins induce enteritis in Atlantic salmon (*Salmo salar* L.). J Agric Food Chem 63:3887–3902

Kumar N (2022) Evaluation of immunostimulatory properties of indigenous medicinal plant *Achyranthes Aspera* in *Labeo rohita* & *Clarias batrachus* in laboratory & pond conditions. Doctoral dissertation

Kumar G, Hummel K, Noebauer K, Welch TJ, Razzazi-Fazeli E, El-Matbouli M (2018) Author correction: proteome analysis reveals a role of rainbow trout lymphoid organs during *Yersinia ruckeri* infection process. Sci Rep 8:15332

Kumar G, Hummel K, Razzazi-Fazeli E, El-Matbouli M (2019a) Modulation of posterior intestinal mucosal proteome in rainbow trout (*Oncorhynchus mykiss*) after *Yersinia ruckeri* infection. Vet Res 50:1–14

Kumar N, Sharma JG, Singh SP, Singh A, Harikrishna V, Chakrabarti R (2019b) Validation of growth enhancing, immunostimulatory and disease resistance properties of *Achyranthes aspera* in *Labeo rohita* fry in pond conditions. Heliyon 5(2):e01246

Kumar N, Sharma J, Kumar G, Shrivastav AK, Tiwari N, Begum A, Chakrabarti R (2021a) Evaluation of nutritional value of prickly chaff flower (*Achyranthes aspera*) as fish feed ingredient. Indian J Anim Sci 91(3):239–244

Kumar G, Sharma J, Goswami RK, Shrivastav AK, Kumar N, Chandra S, Chakrabarti R (2021b) The study of effect of vitamin C and *Achyranthes aspera* seeds enriched diets on the growth, biochemical composition, digestive enzyme activities and expressions of genes involved in the biosynthesis of fatty acids in Snow trout *Schizothorax richardsonii* (Gray, 1832). J Appl Aquac 35:489–509

Kumar G, Sharma J, Goswami RK, Shrivastav AK, Tocher DR, Kumar N, Chakrabarti R (2022a) Freshwater macrophytes: a potential source of minerals and fatty acids for fish, poultry, and livestock. Front Nutr 9:869425

Kumar N, Sharma JG, Mittal P, Chakrabarti R (2022b) Effect of leaves and seeds of *Achyranthes aspera* as feed supplements on the immunological and stress parameters and related-gene expressions of Asian catfish (*Clarias batrachus*). Vet Res Commun 47:99–109

Latosinska A, Vougas K, Makridakis M, Klein J, Mullen W, Abbas M, Stravodimos K, Katafigiotis I, Merseburger AS, Zoidakis J, Mischak H (2015) Comparative analysis of label-free and 8-plex iTRAQ approach for quantitative tissue proteomic analysis. PLoS One 10:e0137048

Liu G, Saxena D, Deng H, Norman RG, Chen Z, Abrams WR, Malamud D, Li Y (2010) Effect of protease inhibitors on the quantitative and qualitative assessment of oral microorganisms. FEMS Microbiol Lett 312:63–70

Ludwig C, Gillet L, Rosenberger G, Amon S, Collins BC, Aebersold R (2018) Data-independent acquisition-based SWATH-MS for quantitative proteomics: a tutorial. Mol Syst Biol 14:e8126

Manes NP, Nita-Lazar A (2018) Application of targeted mass spectrometry in bottom-up proteomics for systems biology research. J Proteome 189:75–90

Mateo-Sagasta J, Zadeh SM, Turral H, Burke J (2017) Water pollution from agriculture: a global review. Executive summary. International Water Management, Rome

Mavi AK, Gour S, Kumar N, Srivastav AK, Bhattacharya S, Belemkar S, Maru S, Kumar D (2023) Effective gene transfer with non-viral vectors. In: Malviya R, Sundram S, Jain N (eds) Integration of biomaterials for gene therapy. Wiley, New York, pp 183–222

McCuaig LM, Martyniuk CJ, Marlatt VL (2020) Morphometric and proteomic responses of early-life stage rainbow trout (*Oncorhynchus mykiss*) to the aquatic herbicide diquat dibromide. Aquat Toxicol 222:105446

Meleady P (2018) Two-dimensional gel electrophoresis and 2D-DIGE. Methods Mol Biol 1664:3–14

Mohanty BP (2015) Proteomics in fisheries and aquaculture. Proteomics Society of India (PSI). PSI Newsletter

Mohanty BP, Mohanty S, Mitra T, Mahanty A, Ganguly S, Singh S (2019) Omics technology in fisheries and aquaculture. In: Advances in fish research, vol 7. Narendra Publishing House, Delhi, pp 1–30

Monroe AA, Zhang H, Schunter C, Ravasi T (2020) Probing SWATH-MS as a tool for proteome level quantification in a non-model fish. Mol Ecol Resour 20(6):1647–1657

Monti M, Cozzolino M, Cozzolino F, Vitiello G, Tedesco R, Flagiello A, Pucci P (2009) Puzzle of protein complexes in vivo: a present and future challenge for functional proteomics. Expert Rev Proteomics 6:159–169

Nynca J, Arnold GJ, Fröhlich T, Ciereszko A (2015) Shotgun proteomics of rainbow trout ovarian fluid. Reprod Fertil Dev 27:504–512

Ortea I, O'Connor G, Maquet A (2016) Review on proteomics for food authentication. J Proteome 147:212–225

Piovesana S, Capriotti AL, Caruso G, Cavaliere C, La Barbera G, Chiozzi RZ, Laganà A (2016) Labeling and label free shotgun proteomics approaches to characterize muscle tissue from farmed and wild gilthead sea bream (*Sparus aurata*). J Chromatogr A 1428:193–201

Piras C, Roncada P, Rodrigues PM, Bonizzi L, Soggiu A (2016) Proteomics in food: quality, safety, microbes, and allergens. Proteomics 16(5):799–815

Power A, Cozzolino D (2020) How fishy is your fish? Authentication, provenance and traceability in fish and seafood by means of vibrational spectroscopy. Appl Sci 10(12):4150

Raposo de Magalhães C, Schrama D, Farinha AP, Revets D, Kuehn A, Planchon S, Rodrigues PM, Cerqueira M (2020) Protein changes as robust signatures of fish chronic stress: a proteomics approach to fish welfare research. BMC Genomics 21:309

Reisinger V, Eichacker LA (2008) Solubilization of membrane protein complexes for blue native PAGE. J Proteome 71(3):277–283

Richard N, Silva TS, Wulff T, Schrama D, Dias JP, Rodrigues PM, Conceição LE (2016) Nutritional mitigation of winter thermal stress in gilthead seabream: associated metabolic pathways and potential indicators of nutritional state. J Proteome 142:1–14

Sakshi, Mavi AK, Srijita C, Kumar N, Singh P, Kumar D, Preethi L, Kumar U (2023) Novel diagnostic methods for emerging respiratory viral infection. In: Bhukya PL, Mhaske ST, Sonkar SC (eds) Emerging human viral diseases, vol 1. Springer, Berlin, pp 565–585

Saleh M, Kumar G, Abdel-Baki AA, Dkhil MA, El-Matbouli M, Al-Quraishy S (2018) Quantitative shotgun proteomics distinguishes wound-healing biomarker signatures in common carp skin mucus in response to *Ichthyophthirius multifiliis*. Vet Res 49:1–12

Schwanhäusser B, Busse D, Li N, Dittmar G, Schuchhardt J, Wolf J, Chen W, Selbach M (2011) Global quantification of mammalian gene expression control. Nature 473:337–342

Semwal A, Kumar A, Kumar N (2023) A review on pathogenicity of *Aeromonas hydrophila* and their mitigation through medicinal herbs in aquaculture. Heliyon 9(3):e14088

Sharma JG, Kumar N, Singh SP, Singh A, HariKrishna V, Chakrabarti R (2019) Evaluation of immunostimulatory properties of prickly chaff flower *Achyranthes aspera* in rohu *Labeo rohita* fry in pond conditions. Aquaculture 505:183–189

Sharma JG, Kumar N, Mittal P, Chakrabarti R (2022) Evaluation of UV-B protective properties of leaves and seeds of *Achyranthes aspera* in Asian catfish *Clarias batrachus* (Linn.). Photochem Photobiol Sci 21:1341–1356

Shi T, Song E, Nie S, Rodland KD, Liu T, Qian WJ, Smith RD (2016) Advances in targeted proteomics and applications to biomedical research. Proteomics 16:2160–2182

Singh R, Chauhan RS, Semwal A, Kumar N (2023a) Efficacy of 17α-methyl testosterone on growth and survival of fry of swordtail, *Xiphophorus helleri*. J Krishi Vigyan 11(2):65–69

Singh R, Semwal A, Kumar N (2023b) Status of coral reefs, their ecological value and conservation management. In: Nimje PP (ed) Modern research and environmental sustainability, vol 3. AkiNik Publications, New Delhi, pp 67–78

Singh R, Shikha, Kumar N, Semwal A (2023c) Anaesthesia in fisheries: implications for welfare. In: Deka B (ed) Zoology and entomology sciences, vol 14. AkiNik Publications, New Delhi, pp 71–89

Singhal N, Kumar M, Kanaujia PK, Virdi JS (2015) MALDI-TOF mass spectrometry: an emerging technology for microbial identification and diagnosis. Front Microbiol 6:791

Stryiński R, Mateos J, Pascual S, González ÁF, Gallardo JM, Łopieńska-Biernat E, Medina I, Carrera M (2019) Proteome profiling of L3 and L4 Anisakis simplex development stages by TMT-based quantitative proteomics. J Proteome 201:1–11

Tomanek L (2011) Environmental proteomics: changes in the proteome of marine organisms in response to environmental stress, pollutants, infection, symbiosis, and development. Annu Rev Mar Sci 3:373–399

Urban PL (2016) Quantitative mass spectrometry: an overview. Philos Trans Math Phys Eng Sci 374:20150382

Vadivel AKA (2015) Gel-based proteomics in plants: time to move on from the tradition. Front Plant Sci 6:369

Vilgis TA (2020) The physics of the mouthfeel of caviar and other fish roe. Int J Gastron Food Sci 19:100192

Volkova PY, Geras'kin SA (2018) 'Omic' technologies as a helpful tool in radioecological research. J Environ Radioact 189:156–167

Vorontsov EA, Rensen E, Prangishvili D, Krupovic M, Chamot-Rooke J (2016) Abundant lysine methylation and N-terminal acetylation in sulfolobus islandicus revealed by bottom-up and top-down proteomics. Mol Cell Proteomics 15:3388–3404

Wu P, Heins ZJ, Muller JT, Katsnelson L, Bruijn I, Abeshouse AA, Schultz N, Fenyö D, Gao J (2019) Integration and analysis of CPTAC proteomics data in the context of cancer genomics in the cBioPortal*[S]. Mol Cell Proteomics 18(9):1893–1898

Xu L, Lu Z, Ji C, Cong M, Li F, Shan X, Wu H (2019) Toxicological effects of as (V) in juvenile rockfish *Sebastes schlegelii* by a combined metabolomic and proteomic approach. Environ Pollut 255:113333

Yadav SP (2007) The wholeness in suffix -omics, -omes, and the word om. J Biomol Tech J Biochem 18(5):277

Yilmaz O, Patinote A, Nguyen TV, Com E, Lavigne R, Pineau C, Sullivan CV, Bobe J (2017) Scrambled eggs: proteomic portraits and novel biomarkers of egg quality in zebrafish (*Danio rerio*). PLoS One 12:e0188084

Yue GH, Wang L (2017) Current status of genome sequencing and its applications in aquaculture. Aquaculture 468:337–347

Zhu L, Gao N, Wang R, Zhang L (2018) Proteomic and metabolomic analysis of marine medaka (*Oryzias melastigma*) after acute ammonia exposure. Ecotoxicology 27:267–277

Nano-biosensor Applications for Water Quality Monitoring

5

Jyotsna Dei, Rajib Bandyopadhyay,
and Bijay Kumar Behera

Abstract

The process of industrialization and urbanization has led to a rise in the quantity of pollutants present in water bodies, degrading the overall quality of water. Both conventional laboratory-based analytical equipment and cutting-edge sensor technologies, such as nano-biosensors, have become more prevalent in order to address the need for regulatory measures in environmental pollution management and early detection. The advancements in nanotechnology and sensors have played a crucial role in detecting environment pollutants. The incorporation of nanotechnology enhances the transduction process, making it more cost-effective, durable, reproducible, and selective. By utilizing a recognition nanomaterial, the biosensor detects an analyte, and a transducer transforms this signal into one that can be understood. This chapter provides a concise overview of the use of nano-biosensors for water quality monitoring in environment.

Keywords

Biosensor · Nanotechnology · Pollution

J. Dei · R. Bandyopadhyay
Department of Instrumentation and Electronics Engineering, Jadavpur University, Kolkata, India

B. K. Behera (✉)
College of Fisheries, Rani Lakshmi Bai Central Agricultural University, Jhansi, Uttar Pradesh, India

B. K. Behera (ed.), *Current Trends in Fisheries Biotechnology*,
https://doi.org/10.1007/978-981-97-3165-7_5

5.1 Introduction

Over the course of history, the assessment of water quality for drinking purposes has been conducted in order to decrease the risk of exposure to disease-causing agents. Surprisingly, over 844 million people lack access to potable water, and over 2 million, mostly children under the age of 5, die annually as a result of inadequate access to clean drinking water (Vikesland 2018). Monitoring the quality of water, whether on a big or small scale, is difficult because of the unpredictability and complexity of the matrix, as well as the low quantities of contaminants that need to be identified (Kumar et al. 2023b). Nano-biosensors are kind of devices that uses nanomaterials as recognition components to identify the changes. Since the first encouraging findings of nanowire and quantum dot-enabled sensing, the research community has increasingly focused on nanomaterial-assisted sensor systems to detect possible substance of their interest. The development of electrochemical biosensors for the detection of environmental contaminants has drawn a lot of interest recently (Gupte and Pradeep 2022). When it comes to evaluating water quality, the biosensors react to the presence of a substance or chemical and send the data to be evaluated by detector. These sensors can be used directly in the center of water treatment plant as well as the source of point. Currently, various nanosensors have been suggested so far for sensitively identifying pH levels, microorganisms, and toxins in water (Hui et al. 2022).

When it comes to nano-biosensors, the incorporation of nanotechnology enhances the transduction process, making it more cost-effective, durable, reproducible, and selective. Nano-biosensors have three components, first one is a nanomaterial, second one is an element for recognition and third one is a mechanism for transduction of signals (Vikesland and Wigginton 2010). By utilizing a recognition nanomaterial, the sensor detects an analyte, and a transducer transforms this signal into one that can be understood. Nanosensor specificity is attained by detecting either an inherent signal sent by the analyte or using recognition components that possess a high degree of specificity, preferably binding exclusively to a particular target. Based on bio-recognition elements such as enzymes, antibodies, receptors, cells, DNA, and molecularly imprinted polymers (MIP) and working principles as well as transducing mechanisms biosensors can be categorized into four types viz. electrochemical, piezoelectric, optical, and thermal biosensors (Turner 2000). This chapter highlights the potential application of nano-biosensors to monitor possible contamination like heavy metals, pesticides, antibiotics, and potential disease-causing microbes.

5.2 Contamination of Water

Water is a crucial need for the life of humans. According to the 2021 World Water Development Report published by UNESCO, the worldwide use of freshwater has multiplied by six times in the last century and has been steadily increasing at a rate of around 1% year since the 1980s (Lin et al. 2022). The environment has been

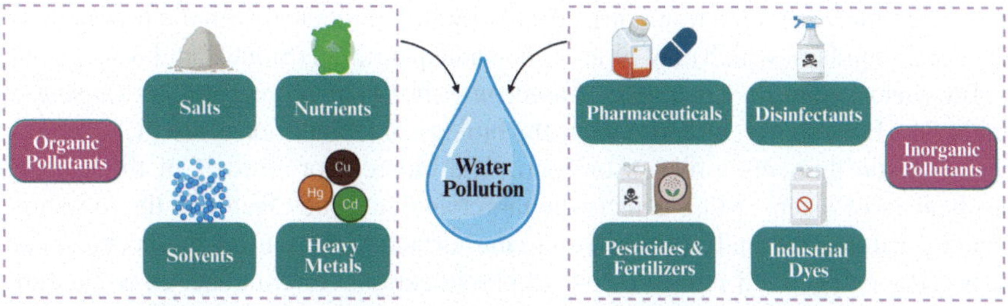

Fig. 5.1 Depiction of organic and inorganic pollutants involved in water pollution

degraded and polluted due to industrialization, agriculture, and urban living. This has harmed the water bodies that support life, human health, and the advancement of sustainable social development (Dhar et al. 2023). A wide range of hazardous pollutants, including poisonous chemicals, organic and inorganic compounds, toxic solvents, and volatile organic chemicals, are released into aquatic ecosystem without proper treatment by various industries causing pollution of water (Chowdhary et al. 2020). Arsenic, cadmium, and chromium are essential contaminants released in wastewater, and the industrial sector plays a key role in the emission of dangerous pollutants. The organic and inorganic sources of water pollution have been outlined in Fig. 5.1. Furthermore, there is a strong correlation between water contamination and agriculture. Agricultural activities, such as the use of pesticides, nitrogen fertilizers, and the disposal of organic agricultural wastes, are major contributors to water contamination (Sharma et al. 2024). Therefore, there is a need to determine whether the specific contaminant exists above the regulatory limit. Nanomaterial-based sensors are a very promising technology that provide exceptional sensitivity in detecting environmental toxins at extremely low concentrations, ranging from nanomolar to sub-picomolar levels. The advantage of these sensors is accuracy to identify contaminants, easy to carry in the field, and eliminating the need for costly laboratory equipment (Mokhtar et al. 2016; Willner and Vikesland 2018).

5.3 Composition and Basic Principle of Biosensors

A biosensor is a compact device consisting of a transducer and a biological component, such as an enzyme, antibody, and nucleic acid used for sensing and detecting biological substances. The bioelement interacts with the analyte and converts the biological reaction into an electrical signal via the transducer. A biosensor consists of a biological component that functions as the sensor and an electrical component that receives and sends the signal. The biological material is fixed in place and brought into contact with the transducer. The analyte attaches to the biological material, resulting in the formation of a bound analyte. This bound analyte generates an electrical response that can be quantified. Occasionally, the analyte undergoes a transformation into a substance that may be linked to the generation of thermal energy, gaseous molecules (such as oxygen), electric charges, or hydrogen ions. The transducer then transforms the alterations associated with the product into electrical impulses, which may be amplified and quantified (Smutok and Katz 2023).

Biosensors may be categorized based on either their bioreceptor or transducer kinds. The bioreceptor, also known as the biological recognition element, is the key characteristic that sets a biosensor apart. The bioreceptor consists of the sensing mechanism of a sensor that detects the target analyte. They facilitate the measurement by affixing the analyte of interest to the surface of the sensor. Bioreceptors can be broadly categorized into five distinct classifications: biomimetic, enzyme, antibody/antigen, nucleic acid/DNA, and cellular structure/cell-based. The sampling component of a biosensor consists of a biosensitive layer, which may either include bioreceptors or be composed of bioreceptors that are covalently bonded to the transducer. The transducer is a vital component of biosensors and plays a crucial part in the process of signal detection. A transducer can transform biorecognition signal events into signals that can be detected (Malik et al. 2023). Electrochemical signals may be detectable via amperometry, conductometry, impedimetry, potentiometry, or voltammetry. The available methods include optical techniques such as colorimetry, fluorescence, luminescence, and interferometry, as well as calorimetric methods using thermistors. The alteration in mass may be induced by either piezoelectric or acoustic means, as well as by magnetic forces. The electrochemical instruments used in conjunction with biosensors have been reduced in size to compact, portable devices that are suitable for both home usage and clinical examination. Electrochemical sensors, including optical and piezoelectric sensors, are the smallest sensors available. This gives electrochemical biosensors the benefit of being portable and easy to use (Karunakaran et al. 2015; Bhalla et al. 2016). Furthermore, the electrochemical sensors have superior sensitivity and responsiveness compared to optical or piezoelectric sensors.

5.4 Detection of Pesticides

Pesticide like insecticides, fungicides, and herbicides are generally used for plant protection. Pesticides can infiltrate the soil and contaminate groundwater, which may lead to drinking water pollution. Additionally, they have the potential to infiltrate surface water via drift or runoff, presenting a hazard to both human well-being and ecosystems (Sjerps et al. 2019; Hara and Singh 2021). Electrochemical biosensors used for pesticide sensing applications include three main types like immunosensors, whole cell-based and enzyme-based. Enzyme-based biosensors have several benefits, such as exceptional specificity, sensitivity, selectivity, and availability. A direct-mode enzymatic biosensor checks the amount of analyte or the product that is made during an enzyme reaction. The performance of these portable instruments in detecting trace-level pesticide residues has been enhanced with the inclusion of innovative nanosensors and biosensors into revolutionary microelectronics-based systems (Mukherjee et al. 2022). An indirect mode biosensor checks how enzymes are inhibited by the analyte of interest. For instance, Sarkar et al. recently developed a biosensor made up of polyvinyl alcohol-capped silver nanoparticle to detect the organophosphate pesticides based on the enzyme acetylcholinesterase (Sarkar et al. 2023). For the purpose of detecting pesticides, electrochemical biosensors including whole microbial cells have also been created as an alternative to enzymes (Gao et al. 2017). In microbial cells, the microbes undergo physiological changes upon exposure to certain toxicants. The changes in respiratory chain activity are recognized on the electrode surface (Tucci et al. 2019). Similarly, electroactive biofilms (EABs) can be integrated into microbial fuel cells (MFCs) and used as self-sustaining sensor devices for the detection of environmental toxins (Jiang et al. 2018; Do et al. 2020). Immunosensor use has several limitations, such as the high expenses associated with antibody synthesis and their susceptibility to instability when employed for on-site detection of environmental toxins (Nguyen et al. 2017).

5.5 Detection of Heavy Metals

In general, heavy metals are characterized as metals with large atomic masses compared to water and high relative densities (>5 g cm^{-3}). Toxic heavy metal pollution of aquatic environments is a global concern due to their resistance to biodegradation, ability to accumulate in sediment and surface water, and potential entry into the food chain. The water contamination with heavy metals results from fertilizer use, mining operations, volcanic eruptions, and industrial activity (Bera et al. 2022). The usage of heavy metals including As, Cu, Cd, Pb, Hg, and Zn may be found in a variety of sectors, including the paint industry, the automobile equipment industry, and the electronics industry (Kladsomboon et al. 2020). Heavy metals harm biological systems through oxidative stress, reduced antioxidant metabolism, and enzyme inhibition. To detect the heavy metals in the environment the biosensor is categorized based on its integrated components. These include transducers and bioreceptors. The bioreceptor-based biosensors to detect heavy metal are further categorized as enzyme-based, antibody-based, whole cell-based, and DNA-based. Furthermore, transducer-based biosensors are categorized as electrochemical-based, optical-based, piezoelectric-based, voltametric-based, and calorimetric-based (Velusamy et al. 2022). Various types of biosensors as well as their use to detect heavy metals are mentioned below.

5.5.1 Enzyme-Based Biosensors for Heavy Metal Detection

Biosensors use immobilized enzymes on the surface of the transducer to serve as a mediator between the analyte and the transducer. Heavy metals often hinder the activity of enzymes. This interaction exhibits high selectivity toward certain ions, resulting in an amplified signal production. Using a thick film interdigitated electrode, Ilangovan et al. created a sol-gel-immobilized-urease biosensor. The biosensor is capable of detecting heavy metal ions in liquid samples. This particular biosensor has been used in order to ascertain the heavy metal ions of varying quantities 0.1–10 mM during time (Ilangovan et al. 2006). New developments in the area point to DNAzyme-based technologies as a potential solution with applications in colorimetric, fluorometric, and conductometric sensitivities. The functionality of some DNAzymes relies on the existence of certain components, such as metal ions and proteins. Due to this characteristic, these DNAzymes may be used to design biosensors that are capable of detecting such molecules (Khan et al. 2021). Functional oligonucleotides, depending on the particular DNAzyme used, may serve as molecular recognition elements in biosensors, providing exceptional selectivity to the target analyte. Alternatively, they can work as reporters that convert a detectable signal. For instance, a Cd^{2+}-fueled DNAzyme waker was developed for the detection of Cd^{2+}. This includes a strong fluorescence signal in the sensing device (Pan et al. 2023).

The process of detecting heavy metals in water using enzyme-based biosensors has been depicted in Fig. 5.2. Antibody is a protein that specifically binds to a certain substance. In this case, it could be designed to bind to a heavy metal or to a metal-induced modification on the bioreceptor. Bioreceptor biological components, such as an enzyme, that has a high affinity for the target analyte, are the heavy metals in this context. The bioreceptor is immobilized on a solid surface to interact with the heavy metals in the water sample. Heavy metals are the target contaminants in the water that need to be detected. They can include metals like lead, cadmium, mercury, etc., which are known to be toxic even at low concentrations. When heavy metals are present in the water sample, they bind to the bioreceptor. This interaction is specific and leads to a measurable change, which can be an alteration in the bioreceptor's activity or structure. After this the transducer converts the biochemical signal resulting from the interaction between the heavy metals and the bioreceptor

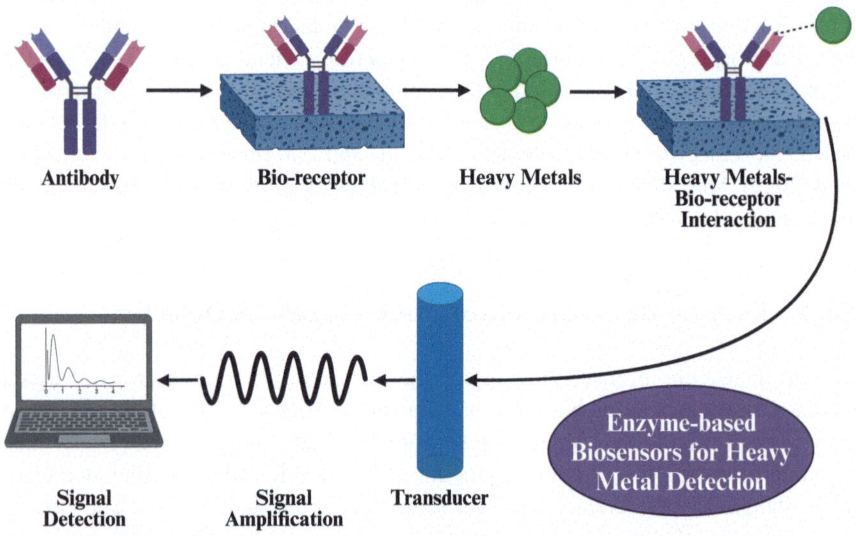

Fig. 5.2 The process of detecting heavy metals in water using enzyme-based biosensors

into a measurable physical signal, often electrical. Then the signal amplification increases the intensity of the physical signal to make it more detectable and quantifiable. Signal amplification is crucial because the interactions between the heavy metals and the bioreceptors can be minute and difficult to detect without amplification. The amplified signal is then detected and recorded by an instrument. This data can be displayed as a graph or readout on a computer, indicating the presence and possibly the concentration of heavy metals in the water sample.

5.5.2 Aptamers-Based Biosensors for Heavy Metal Detection

Aptamers are categorized as either short-chain nucleotides or oligos that are chemically produced using the binding affinity of heavy metals. The oligos are typically manufactured using the sequential systematic evolution of ligands by exponential enrichment (SELEX) approach. The fundamental concept underlying aptamer-based sensing can be categorized into two types: first, by immobilizing the aptamer on the surface of an electrochemical electrode to enhance sensitivity, and second, by utilizing aptamer-based substrates like gold nanoparticles (AuNP) to induce a colorimetric response (Mukherjee et al. 2021). Both of these methods may be used to determine the particular metals contained in a sample. Their sensitivity and limit of detection (LOD) are often not as minimal as those of traditional methods, but they are very efficient and have a high throughput in detecting contaminated materials. Wang and Si have used a new aptamer biosensor that relies on multi-walled carbon nanotubes (MWCNT). This biosensor enables fluorescence energy transfer when it interacts with heavy metals like Pb^{2+} ions, facilitating their swift detection up to 20 nM (Wang and Si 2013). Similarly, Peng et al. design a novel aptamer-peptide-based conjugate to gain an enhanced affinity toward Pb^{2+} which suggests the increased affinity by 62.96% and 802.56% for aptamers and peptides (Peng et al. 2023). In addition, Zhou et al. have described a technique employing click chemistry to detect Cu^{2+} ions in aqueous solutions using gold nanoparticles that are functionalized with azide and terminal alkyne groups. The catalyst, Cu(I), was simply obtained by the reduction of Cu(II) in the presence of sodium ascorbate. This technique enables direct observation of Cu^{2+} ions via the clustering of AuNPs caused by the Cu(I)-catalyzed conjugation between the two functional groups, without the need for any sophisticated equipment (Zhou et al. 2008).

In an example as shown in Fig. 5.3, there can be two scenarios in the detection of heavy metals in water using aptamer-based biosensors. Aptamers are short, single-stranded DNA or RNA molecules that can bind selectively to specific targets, including heavy metal ions. The presence of a heavy metal ion, exemplified by mercury, induces a conformational change in the DNA aptamer structure. This change brings two components closer together: DNA1 tagged with a gold nanoparticle and DNA2 attached to fluorescence microspheres. When these components are in proximity due to the structural change caused by the metal ion, a fluorescent signal is emitted, indicating the presence of the heavy metal. In the absence of the heavy metal ion the DNA aptamers do not undergo a conformational change and

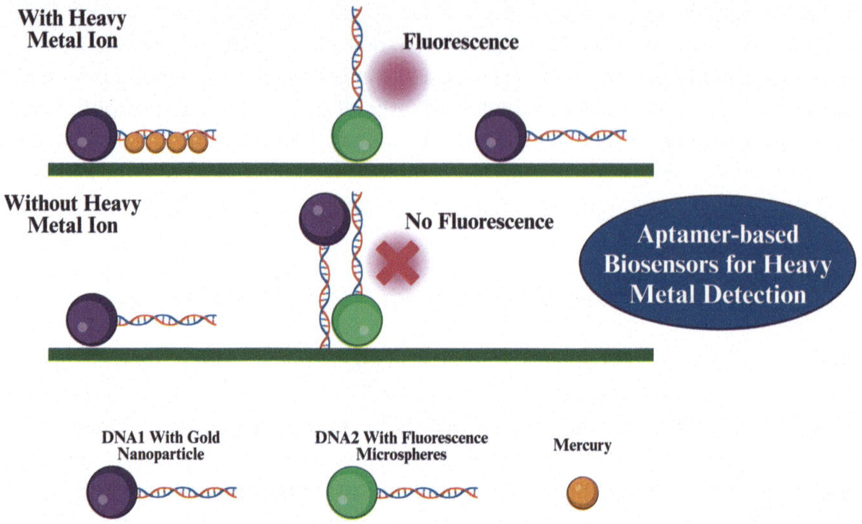

Fig. 5.3 The process of detecting heavy metals in water using aptamer-based biosensors

thus the gold nanoparticle (DNA1) and the fluorescence microsphere (DNA2) remain separate (Berlina et al. 2019). This separation prevents the occurrence of fluorescence, signaling that no heavy metal ions are detected.

5.5.3 Ion-Imprinted Sensors for Heavy Metal Detection

Ion-imprinted polymers possess distinct cavities that are designed to selectively bind to various ions. Typically, polymers are formed by the process of polymerization, when monomers react in the presence of ions. The ions are removed in a sequential manner from the polymer, resulting in the formation of corresponding cavities within the polymer matrix. The polymer matrix forms cavities that selectively bind certain ions, generating signals that may be recorded and analyzed. The electrodes can be readily polymerized with these ion-imprinted polymers in order to conduct electrochemical detection of analyte molecules or molecular sensors (Mukherjee et al. 2022). For instance, Zhang et al. developed a biosensor based on recognition strategy using an ion-imprinted polymer and Bovine serum albumin (BSA). The gold electrode surface was modified using bovine serum albumin (BSA) and graphene oxide (GO)-loaded gold nanoparticles (GO/Au NPs). Subsequently, a Cu^{2+} ion-imprinted polymer (IIP) was created on the electrode by electropolymerization with a detection limit of 1.12×10^{-15} mol/L (Zhang et al. 2023). The list of biosensors used for heavy metal detection has been mentioned in Table 5.1.

Table 5.1 List of biosensors used for heavy metal detection

Sl. no	Detection	Type of sensors	LOD	References
1.	Ag^{+}, Pb^{2+}, Cu^{2+}	Horseradish peroxidase (HRP) immobilized within oxidized PSi nanostructure	60–120 ppb	Shtenberg et al. (2015)
2.	Cu^{2+}, Hg^{2+}, Zn^{2+}, Cd^{2+}, Pb^{2+}, Cr^{3+}	Microbial fuel cell (MFC)-based	2 mg/L	Yu et al. (2017)
3.	Hg^{2+}	The fluorescence-labeled DNA will attach to the DNA probes that are fixed onto the surface of the sensor (optical biosensor)	1.2 nM	Long et al. (2013)
4.	Cu^{2+} and Al	CQD-loaded chitosan-based thin-film biosensors	0–100 μM	Vyas et al. (2023)
5.	Hg^{2+}	Double-Hemin G-Quadruplex	0.1 nM	Bao et al. (2023)
6.	Cd^{2+}	CRISPR/Cas12a with aptamers	1.25 pM	Liang et al. (2023)
7.	Cu^{2+}	DNAzyme-based fluorescence biosensor	60 fM	Liu et al. (2023)
8.	As(III)	Aptamer (QCM)	8.6 nM	Kumar et al. (2023a)
9.	Cd^{2+}	DNAzyme and stem-loop label-free qPCR	6 nmol L^{-1}	Li et al. (2023)
10.	Pb^{2+}, Cd^{2+}, Cr^{3+}	DNAzyme electrochemical biosensor	20 nM, 20 nM, 44 nm	Skotadis et al. (2023)
11.	Pb^{2+}	DNAzyme on solution-gated graphene transistor (SGGT)	1.9 nM	Zhao et al. (2023)
12.	Pb(II)	Whole-cell proviolacein (PV)-based	0.183 nM	Zhu et al. (2023)
13.	Cu^{2+}, Pb^{2+}	Liquid crystal (LC)-based biosensors	1 nM	Ul Amin et al. (2020)
14.	Cd^{2+}, Pb^{2+}	Carbon-based sensor (μCS)	1.2, 1.8 μg/L	Shen et al. (2017)
15.	Cr(VI)	3,3′,5,5′-tetramethylbenzidine (TMB)-based sensing probe	25–600 ppb	Dei et al. (2024)

5.6 Detection of Antibiotics

Antibiotics are frequently employed in managing and preventing infectious diseases in animals due to their bactericidal properties. The presence of residual antibiotics poses significant risks to human and animal health due to possible carcinogenic and mutagenesis effects and the development of anti-bacterial resistance. In recent decades, the use of biosensors increases sensitivity, reaction time, and affordable cost to identify antibiotic residue in the environment. In particular applications, such as controlling commercial pharmaceutical specialty, it is necessary to combine antibodies with signal amplification in order to achieve increased sensitivity (Zhou et al. 2021). This is because antibiotics often have low molecular weights. For example, chloramphenicol, which has a molecular weight of 323.13, is considered a hapten which is incapable of using in biosensors. The diazotization procedure is used to prepare its immunogenicity. Chloramphenicol-protein conjugates are used as capture probes in biosensors to detect chloramphenicol (Li et al. 2020). Another approach to detect antibiotics is aptamer-based biosensors. The aptamers are single-stranded DNA that is easy to synthesize with phenomenal chemical stability making them ideal sensing elements used as a sensing material. Aptamers adhere to the surface of gold nanoparticles (AuNPs) via electrostatic interactions and it will only detach the surface upon binding with its target element. For instance, Seo et al. introduced a colorimetric aptasensor that utilizes AuNPs to detect oxytetracycline based on reflectance. This method has the advantages of straightforward operation and visual detection without the need for specialized equipment (Seo et al. 2015). Recently the use of virus has also gained popularity due to its surface affinity towards various chemical recognition elements. For example, modified *Tobacco mosaic virus* particle functionalizing with recognition molecules allows for seamless integration with various transducers, hence creating novel possibilities for biosensing for various antibiotics detection (Poghossian et al. 2018). Currently, antibiotics may be detected using whole-cell biosensors, which is one of the promising antibiotic detection methods. The study suggested that the biosensors for the antibiotics target peptidoglycan (PG) layer of the bacterial cell wall. For instance, a very precise and accurate biosensor has been developed using a unique two-component system found in the Gram-negative bacteria *Shewanella oneidensis*. This biosensor is designed to detect and react to PG stress enabling them to identify antibiotics in environment (Yin et al. 2022).

5.7 Detection of Pathogens

Historically, the identification of bacterial pathogens has been accomplished by the use of several methodologies that depend on cultivating the pathogens using different types of selective medium (Kumar et al. 2022; Behera et al. 2023). Although the conventional methods of pathogen detection have proven effective in diagnosing numerous pathogenic bacteria in food, water, and the environment, they are time-consuming and require a growth period of approximately 18–72 h to achieve a measurable pathogen concentration (Valenzuela-Amaro et al. 2023). Additionally, some pathogens like *Vibrio cholerae* and *Helicobacter pylori* may exist in a food, drink, or environmental sample in a condition known as viable but not culturable (VBNC). As a result, they cannot be cultivated using standard growth medium. These challenges have increased the need for a more rapid method of pathogen detection to safeguard human health (Nnachi et al. 2022). The use of biosensors provides significant advantages over traditional methods in detecting bacterial pathogens in food, water, and the environment due to high sensitivity reduced labor intensity, and high effectiveness (Paul et al. 2019). Whole-cell biosensors developed using synthetic biology have significant potential for the quick and cost-efficient identification of waterborne infections. The biosensor identifies the specific bacterial pathogens by detecting and reacting to the corresponding quorum-sensing signal molecules. For instance, a biosensor using the QscR quorum-sensing signal system was developed by Wu et al. to detect the two most bacterial pathogens *Burkholderia pseudomallei* and *Pseudomonas aeruginosa*. Aptamers function as a molecular biorecognition component in aptasensors. They have enhanced target recognition and binding capabilities throughout a wide range of concentrations, from extremely low level, when they adopt their secondary and tertiary structures (Rahimizadeh et al. 2023). Similarly, extremely low concentration of pathogens can be detected by using SERS microfluidic biosensors. For instance, pathogenic *Escherichia coli* O157:H7 can be detected at a minimum concentration of 10 CFU/mL by using a nano-dielectrophoretic microfluidic device combined with the surface enhanced Raman scattering (SERS) approach (Wang et al. 2017). Another approach to detect pathogenic bacteria is the use of triboelectric nanogenerators (TENGs). TENGs, also known as triboelectric nanogenerators, are a type of biosensors that efficiently transform mechanical energy into electrical energy, which may be used to operate low-power electronic devices. By utilizing this sensor system, a platform consisting of 3D printed polylactic acid (PLA) functionalized with D-mannose and silver electrodes painted with brushes is used to capture pathogenic *E. coli* bacteria. This capture is achieved by a straightforward process involving the interaction between a simple carbohydrate and protein with biodegradability activity (Panda et al. 2023).

5.8 Conclusion

Despite the excellent sensitivity and cost-effectiveness of biosensors, there is a need to enhance their performance as well as the ability to detect multiple pollutants simultaneously. Novel nanocomposites and nanomaterials, together with innovative functionalization processes, will be used to enhance the biosensors designed for environmental quality monitoring. The successful incorporation of nanomaterials with biomolecules onto the surface of electrodes or nanofilms has the potential to give rise to a new and innovative generation of biosensors which can be used to monitor contaminants in water. Another approach is to use AI-based biosensor framework for automatic use of sensors for accurate detection of pathogens without constant efforts by human. Although each type of biosensor is highly cost-effective and sensitive, there still needs to improve their performance and utility toward various types of water contaminants at the same time point. However, the use of biosensors on a commercial scale is still restricted owing to a number of issues, including the size, cost, detection range, and leakage of biological components. Despite significant progress, there is still a great deal to be achieved and explored in this area.

References

Bao W, Ga L, Ai J (2023) Double-hemin G-quadruplex highly sensitive detection of mercury ions in human serum

Behera BK, Parida SN, Kumar V et al (2023) *Aeromonas veronii* is a lethal pathogen isolated from gut of infected Labeo rohita: molecular insight to understand the bacterial virulence and its induced host immunity. Pathogens 12:598

Bera T, Kumar V, Sarkar DJ et al (2022) Pollution assessment and mapping of potentially toxic elements (PTE) distribution in urban wastewater fed natural wetland, Kolkata, India. Environ Sci Pollut Res 29(45):67801–67820

Berlina AN, Zherdev AV, Dzantiev BB (2019) Progress in rapid optical assays for heavy metal ions based on the use of nanoparticles and receptor molecules. Microchim Acta 186:1–39

Bhalla N, Jolly P, Formisano N, Estrela P (2016) Introduction to biosensors. Essays Biochem 60:1–8

Chowdhary P, Bharagava RN, Mishra S, Khan N (2020) Role of industries in water scarcity and its adverse effects on environment and human health. Environ Concerns Sustain Dev 2029:235–256

Dei J, Mondal S, Biswas A et al (2024) Cr-detector: a simple chemosensing system for onsite Cr (VI) detection in water. PLoS One 19:e0295687

Dhar S, Devnath S, Kumar V et al (2023) Bioremediation and its application in aquaculture. In: Biotechnological tools in fisheries and aquatic health management. Springer, Berlin, pp 197–215

Do MH, Ngo HH, Guo W et al (2020) Microbial fuel cell-based biosensor for online monitoring wastewater quality: a critical review. Sci Total Environ 712:135612

Gao G, Fang D, Yu Y et al (2017) A double-mediator based whole cell electrochemical biosensor for acute biotoxicity assessment of wastewater. Talanta 167:208–216

Gupte T, Pradeep T (2022) Nanosensors for water quality monitoring. Sep Sci Technol 15:37–53

Hara TO, Singh B (2021) Electrochemical biosensors for detection of pesticides and heavy metal toxicants in water: recent trends and progress. ACS Environ Sci Technol Water 1:462–478

Hui Y, Huang Z, Alahi MEE et al (2022) Recent advancements in electrochemical biosensors for monitoring the water quality. Biosensors 12:551

Ilangovan R, Daniel D, Krastanov A et al (2006) Enzyme based biosensor for heavy metal ions determination. Biotechnol Biotechnol Equip 20:184–189

Jiang Y, Yang X, Liang P et al (2018) Microbial fuel cell sensors for water quality early warning systems: fundamentals, signal resolution, optimization and future challenges. Renew Sust Energ Rev 81:292–305

Karunakaran C, Rajkumar R, Bhargava K (2015) Introduction to biosensors. Biosens Bioelectron:1–68

Khan S, Burciu B, Filipe CDM et al (2021) DNAzyme-based biosensors: immobilization strategies, applications, and future prospective. ACS Nano 15:13943–13969

Kladsomboon S, Jaiyen C, Choprathumma C et al (2020) Heavy metals contamination in soil, surface water, crops, and resident blood in Uthai District, Phra Nakhon Si Ayutthaya, Thailand. Environ Geochem Health 42:545–561

Kumar V, Das BK, Swain HS et al (2022) Outbreak of Ichthyophthirius multifiliis associated with Aeromonas hydrophila in Pangasianodon hypophthalmus: the role of turmeric oil in enhancing immunity and inducing resistance against co-infection. Front Immunol 13:956478

Kumar S, Singh P, Verma SK (2023a) Selection of highly specific DNA aptamer for the development of QCM-based arsenic sensor. Chembiochem 24:e202300585

Kumar V, Parida SN, Roy S et al (2023b) An overview of modern biotechnological tools in aquatic food production. Ann Anim Sci 24(1):13–25

Li P, Yu J, Zhao K et al (2020) Efficient enhancement of electrochemiluminescence from tin disulfide quantum dots by hollow titanium dioxide spherical shell for highly sensitive detection of chloramphenicol. Biosens Bioelectron 147:111790

Li J, Du Z, Wang P et al (2023) A turn-on signal biosensor for cadmium(II) based on DNAzyme and stem-loop qPCR. Anal Chim Acta 1279:341827

Liang P, Lv B, Chen K et al (2023) An ultrasensitive Cd2+ detection biosensor based on DNAzyme and CRISPR/Cas12a coupled with hybridization chain reaction. Anal Chim Acta 1283:341950

Lin L, Yang H, Xu X (2022) Effects of water pollution on human health and disease heterogeneity: a review. Front Environ Sci 10:880246

Liu Z, Liu C, He L et al (2023) A cascade signal amplification strategy for the ultrasensitive fluorescence detection of Cu2+ via λ-exonuclease-assisted target recycling with mismatched catalytic hairpin assembly. Biosensors (Basel) 13:918

Long F, Zhu A, Shi H et al (2013) Rapid on-site/in-situ detection of heavy metal ions in environmental water using a structure-switching DNA optical biosensor. Sci Rep 3:1–7

Malik S, Singh J, Goyat R et al (2023) Nanomaterials-based biosensor and their applications: a review. Heliyon 9:e19929

Mokhtar B, Azab M, Shehata N, Rizk M (2016) System-aware smart network management for nano-enriched water quality monitoring. J Sens 2016:3023018

Mukherjee S, Bhattacharyya S, Ghosh K et al (2021) Sensory development for heavy metal detection: a review on translation from conventional analysis to field-portable sensor. Trends Food Sci Technol 109:674–689

Mukherjee S, Ghosh K, Bhattacharyya S et al (2022) A review on recent trends in advancement of bio-sensory techniques toward pesticide detection. Food Anal Methods 15(12):3416–3434

Nguyen VT, Kwon YS, Gu MB (2017) Aptamer-based environmental biosensors for small molecule contaminants. Curr Opin Biotechnol 45:15–23

Nnachi RC, Sui N, Ke B et al (2022) Biosensors for rapid detection of bacterial pathogens in water, food and environment. Environ Int 166:107357

Pan J, Deng F, Chen J (2023) A fluorescent biosensor for Cd2+ detection in water samples based on Cd2+-fueled wheel DNAzyme walker and its logic gate applications. Sci Total Environ 864:161046

Panda S, Hajra S, Kim HG et al (2023) Carbohydrate–protein interaction-based detection of pathogenic bacteria using a biodegradable self-powered biosensor. J Mater Chem B 11:10147–10157

Paul R, Saville AC, Hansel JC et al (2019) Extraction of plant DNA by microneedle patch for rapid detection of plant diseases. ACS Nano 13:6540

Peng K, Liu X, Yuan H et al (2023) A novel fluorescent biosensor based on affinity-enhanced aptamer-peptide conjugate for sensitive detection of lead(II) in aquatic products. Anal Bioanal Chem 415:3463–3474

Poghossian A, Jablonski M, Koch C et al (2018) Field-effect biosensor using virus particles as scaffolds for enzyme immobilization. Biosens Bioelectron 110:168–174

Rahimizadeh K, Zahra QUA, Chen S et al (2023) Nanoparticles-assisted aptamer biosensing for the detection of environmental pathogens. Environ Res 238:117123

Sarkar DJ, Goswami S, Sahu S et al (2023) An amperometric acetylcholinesterase biosensor based on polyvinyl-alcohol-capped silver nanoparticles (AgNPs@PVA) for detection of organophosphate pesticides. ChemistrySelect 8:e202300688

Seo HB, Kwon YS, Lee JE et al (2015) A novel reflectance-based aptasensor using gold nanoparticles for the detection of oxytetracycline. Analyst 140:6671–6675

Sharma K, Rajan S, Nayak SK (2024) Water pollution: primary sources and associated human health hazards with special emphasis on rural areas. In: Water resources management for rural development. Elsevier, Amsterdam, pp 3–14

Shen LL, Zhang GR, Li W et al (2017) Modifier-free microfluidic electrochemical sensor for heavy-metal detection. ACS Omega 2:4593–4603

Shtenberg G, Massad-Ivanir N, Segal E (2015) Detection of trace heavy metal ions in water by nanostructured porous Si biosensors. Analyst 140:4507–4514

Sjerps RMA, Kooij PJF, van Loon A, Van Wezel AP (2019) Occurrence of pesticides in Dutch drinking water sources. Chemosphere 235:510–518

Skotadis E, Aslanidis E, Tsekenis G et al (2023) Hybrid nanoparticle/DNAzyme electrochemical biosensor for the detection of divalent heavy metal ions and Cr3+. Sensors 23:7818

Smutok O, Katz E (2023) Biosensors: electrochemical devices—general concepts and performance. Biosensors 13(1):44

Tucci M, Grattieri M, Schievano A et al (2019) Microbial amperometric biosensor for online herbicide detection: photocurrent inhibition of Anabaena variabilis. Electrochim Acta 302:102–108

Turner APF (2000) Biosensors—sense and sensitivity. Science 290:1315–1317

Ul Amin N, Siddiqi HM, Lin YK et al (2020) Bovine serum albumin protein-based liquid crystal biosensors for optical detection of toxic heavy metals in water. Sensors 20(1):298

Valenzuela-Amaro HM, Aguayo-Acosta A, Meléndez-Sánchez ER et al (2023) Emerging applications of nanobiosensors in pathogen detection in water and food. Biosensors 13(10):922

Velusamy K, Periyasamy S, Kumar PS et al (2022) Biosensor for heavy metals detection in wastewater: a review. Food Chem Toxicol 168:113307

Vikesland PJ (2018) Nanosensors for water quality monitoring. Nat Nanotechnol 13(8):651–660

Vikesland PJ, Wigginton KR (2010) Nanomaterial enabled biosensors for pathogen monitoring—a review. Environ Sci Technol 44:3656–3669

Vyas T, Choudhary S, Sharan Rathnam S, Joshi A (2023) Fiber-optic detection of aluminum and copper in real water samples using enzyme-carbon quantum dot (CQD)-based thin-film biosensors. ACS EST Eng 4:694

Wang SE, Si S (2013) Aptamer biosensing platform based on carbon nanotube long-range energy transfer for sensitive, selective and multicolor fluorescent heavy metal ion analysis. Anal Methods 5:2947–2953

Wang C, Madiyar F, Yu C, Li J (2017) Detection of extremely low concentration waterborne pathogen using a multiplexing self-referencing SERS microfluidic biosensor. J Biol Eng 11:1–11

Willner MR, Vikesland PJ (2018) Nanomaterial enabled sensors for environmental contaminants. J Nanobiotechnol 16(1):95

Yin J, Zhu Y, Liang Y et al (2022) Development of whole-cell biosensors for screening of peptidoglycan-targeting antibiotics in a gram-negative bacterium. Appl Environ Microbiol 88:e0084622

Yu D, Bai L, Zhai J et al (2017) Toxicity detection in water containing heavy metal ions with a self-powered microbial fuel cell-based biosensor. Talanta 168:210–216

Zhang L, Li D, Li S et al (2023) High-sensitivity copper(II) sensor based on a protein/ion-imprinted polymer cooperative recognition strategy. Microchem J 185:108285

Zhao S, Yang J, Wang L et al (2023) Selective detection of Pb2+ ions based on a graphene field-effect transistor gated by DNAzymes in binding mode. Biosens Bioelectron 237:115549

Zhou Y, Wang S, Zhang K, Jiang X (2008) Visual detection of copper(II) by azide- and alkyne-functionalized gold nanoparticles using click chemistry. Angew Chem Int Ed 47:7454–7456

Zhou C, Zou H, Sun C, Li Y (2021) Recent advances in biosensors for antibiotic detection: selectivity and signal amplification with nanomaterials. Food Chem 361:130109

Zhu DL, Guo Y, Ma BC et al (2023) Pb(II)-inducible proviolacein biosynthesis enables a dual-color biosensor toward environmental lead. Front Microbiol 14:1218933

6 Microbial Ecology in Microplastics: Impact on Aquatic Ecosystems and Bioremediation

Sujata Dey, Ajaya Kumar Rout, Koushik Ghosh, Asim Kumar Jana, and Bijay Kumar Behera

Abstract

Microplastics (MPs) and nanoplastics (NPs), which are hazardous contaminants that impair the environment, biota, and human health, are made from plastic that has been dumped into the environment. These pollutants are widespread environmental hazards with negative effects on human health. Reactive oxygen species (ROS) generation in aquatic environments facilitates the photodegradation of MPs into NPs. The photodegradation mechanism of MPs into NPs is triggered by UV radiation. The process is then directed by free radicals (hydroxyl, alkoxyl, and alkyl peroxyl) and reactive oxygen species (ROS), which start a chain reaction and eventually convert MPs to NPs. Sedimentation and buoyancy are important factors in the conversion of MPs into NPs. One problem posed by plastic particles is the surface adsorption of dangerous chemical pollutants, which may have an impact on how they enter the environment. MPs and NPs serve as vectors

S. Dey
Aquatic Environmental Biotechnology and Nanotechnology Division, ICAR-Central Inland Fisheries Research Institute, Barrackpore, Kolkata, West Bengal, India

Aquaculture Laboratory, Department of Zoology, The University of Burdwan, Burdwan, West Bengal, India

A. K. Rout · A. K. Jana
Aquatic Environmental Biotechnology and Nanotechnology Division, ICAR-Central Inland Fisheries Research Institute, Barrackpore, Kolkata, West Bengal, India

K. Ghosh
Aquaculture Laboratory, Department of Zoology, The University of Burdwan, Burdwan, West Bengal, India

B. K. Behera (✉)
College of Fisheries, Rani Lakshmi Bai Central Agricultural University, Jhansi, Uttar Pradesh, India

B. K. Behera (ed.), *Current Trends in Fisheries Biotechnology*,
https://doi.org/10.1007/978-981-97-3165-7_6

for contamination by transporting pollutants (metallic, organic, etc.) and exposing the biota to toxicity. Additionally, tiny plastic particles can carry dangerous microorganisms. For example, MP surfaces can adsorb microbes that are resistant to a variety of antibiotics. Advances in polymer science have been examined concerning the toxicity, transport, and fate of MPs and NPs in earlier studies. This chapter discusses how biofilms affect genes and ecosystems in the environment, as well as how they affect plastic surfaces (such as the plastisphere). It also goes through how the chemistry and ecotoxicity of plastics can be impacted by interactions between bacteria and microplastics. The need for regulation in reducing plastic pollution is explored, along with biodegradation, phytoremediation, and other remediation approaches for MPs and NPs.

Keywords

Microplastics · Biofilms · Microbial ecology · Ecological impacts · Bioremediation

6.1 Introduction

The IUPAC defines plastics as "polymeric materials that may contain additional chemicals to boost performance and/or cut prices" (Chang et al. 2020). Over the previous five decades, the output of plastic waste has expanded at an annual pace of 8.7%, reaching 9.1 billion tons today (UNEP 2016). Microplastics (MPs) and nanoplastics (NPs), which are hazardous contaminants that impair the environment, biota, and human health, are made from plastic that has been dumped into the environment (Rochman et al. 2013). These pollutants are widespread environmental hazards with negative effects on human health (Shen et al. 2020). MPs are plastic particles that measure 5 mm (Masura et al. 2015). NPs are particles with a minimum size of 100 nm in at least one dimension (Lambert and Wagner 2016). Reactive oxygen species (ROS) generation in aquatic environments facilitates the photodegradation of MPs into NPs (Zhu et al. 2020a, b). The photodegradation mechanism here is triggered by UV radiation. The process is then directed by free radicals (hydroxyl, alkoxyl, and alkyl peroxyl) and reactive oxygen species (ROS), which start a chain reaction and eventually convert MPs to NPs (Wang et al. 2021) (Fig. 6.1). Sedimentation and buoyancy are important factors in the conversion of MPs into NPs (Gigault et al. 2018). According to the features of nanoscale plastic growth, these materials behave colloidally like natural particles. For pollutants (such as organic and inorganic), the colloidal property impacts chemical partitioning as well as sorption mechanisms (Wang et al. 2021).

The significant ecological impacts of MPs and NPs, such as changes in ecosystem health, biogeochemical cycling, carbon sequestration, and microbial community dynamics are being understood by researchers (Shen et al. 2019). A key source of plastic particles in aquatic habitats is wastewater treatment plants (WWTPs), which also produce other household plastic waste, synthetic microfibers from laundry, and cosmetics (Alvim et al. 2020; van den Berg et al. 2020). It is possible to introduce plastic particles to terrestrial ecosystems by using irrigation water from

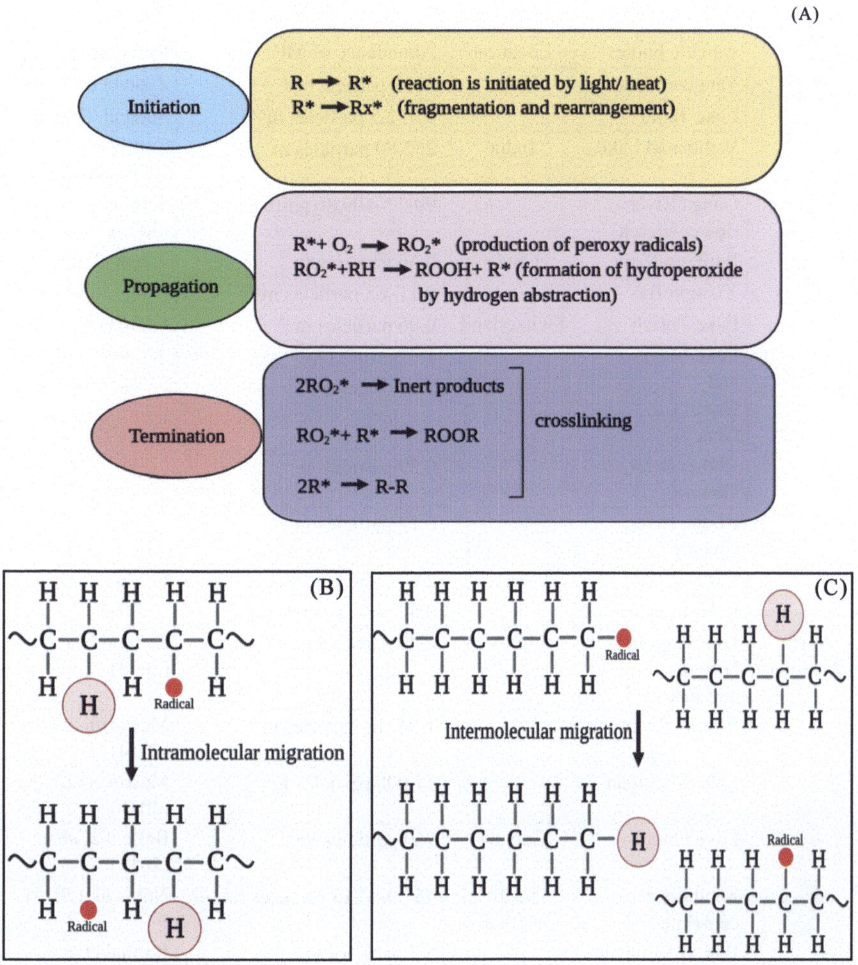

Fig. 6.1 Degradation of polymers through common radical reactions (Inderthal et al. 2021). (**a**) Light and heat facilitate the initiation of the auto-oxidatiosn process followed by propagation and termination. (**b**) Intramolecular and (**c**) intermolecular hydrogen transfer reaction involves the elimination and exchange of hydrogen atoms

agricultural systems and soil amendment derived from MPs/NPs-contaminated WWTP effluent (Ren et al. 2020). According to estimations, farmlands in North America (44,000–300,000 tons annually) and Europe (63,000–430,000 tons annually) are contaminated by the usage of sewage water and sludge (Ng et al. 2018). Soil conditioners, mulching film made of plastic, and greenhouse chemicals are further potential contaminants (de Souza Machado et al. 2018; Huang et al. 2020). Freshwater, marine, and terrestrial ecosystems are all affected by MP and NP pollution (Strungaru et al. 2019) (Table 6.1). Even now, these contaminants are being identified in the atmosphere by new investigations (Zhang et al. 2020a). MP

Table 6.1 Worldwide report of microplastic (MP) abundance in aquatic environments

	Aquatic bodies	Location	Abundance of MP	References
Asia	Yangtze Estuary	China	4137 particle m^{-3}	Zhao et al. (2014)
	Lake Taihu		0.3–2.5 particles m^{-3}	Su et al. (2016)
	Vembanad Lake	India	252.80 particles m^{-2}	Sruthy and Ramasamy (2017)
	Ganga River (lower stretch)		99.27–409.86 particles kg^{-1}	Sarkar et al. (2019)
	Beijiang River	China	0.56 particles m^{-3}	Tan et al. (2019)
	Xiangxi Bay		0.11–68 particles m^{-3}	
Europe	Lake Zurich	Switzerland	0.06 particles m^{-3}	Faure et al. (2013)
	Lake Maggiore and Grand Lac Geneva		1.2 particles m^{-3}	
	Rhone River, Geneva		0.29 particles m^{-3}	
	Rhone River, Chancy		0.13 particles m^{-3}	
	Lake Chiusi	Italy	2.68–3.36 particles m^{-3}	Fischer et al. (2016)
	Lake Bolsena		0.82–4.42 particles m^{-3}	
North America	Lake Superior, Huron, and Erie	USA	0.27 particles m^{-3}	Eriksen et al. (2013)
	Illinois River		1.94–18 particles m^{-3}	McCormick et al. (2014)
	Lake Michigan		17,000 particles km^{-2}	Mason et al. (2016)
	Lake Ontario	Canada	760 particles kg^{-1}	Ballent et al. (2016)
Africa	South-eastern coastline	South Africa	257.9–1215 particles m^{-3}	Nel et al. (2015)
	Bay of Biscay to Cape Town		8.5 particles m^{-3}	Kanhai et al. (2017)
	Bloukrans River		6.3–160.1 particles kg^{-1}	Nel et al. (2018)

pollution was considered among the rising worldwide environmental issues by the United Nations Environment Program (UNEP 2014). One problem posed by plastic particles is the surface adsorption of dangerous chemical pollutants, which may have an impact on how they enter the environment (Ribeiro et al. 2019; Rochman et al. 2014). They act as contamination vectors by transporting pollutants (metallic, organic, etc.) and exposing the biota to toxicity. Additionally, tiny plastic particles can carry dangerous microorganisms (Zhang et al. 2020a, b, c). They can adsorb microbes that are resistant to a variety of antibiotics (Zhang et al. 2020c). Advances in polymer science have been examined concerning the toxicity, transport, and fate of plastic particles in earlier studies (Kershaw and Rochman 2015; Zhu et al., 2020b). This chapter discusses how biofilms affect genes and ecosystems in the environment, as well as how they affect plastic surfaces (such as the plastisphere). It also goes through how the chemistry and ecotoxicity of plastics can be impacted by interactions between bacteria and microplastics. The need for regulation in reducing plastic pollution is explored, along with biodegradation, phytoremediation, and other remediation approaches for MPs and NPs. The management of plastic wastes should be developed to successfully deploy remediation solutions based on different principles (biological, biotechnological, physical, and chemical) in combination with legal regulatory procedures.

6.2 Biofilms: Microplastic-Microbe Interaction in Aquatic Ecosystems

The fate and ecotoxicity of chemical pollutants are influenced by complex, reciprocal interactions in aquatic environments between MPs and microorganisms (micro-with-micro interactions) (Arias-Andres et al. 2019). The term "plastisphere" refers to the particular niche developed around MPs that houses microbial communities or microbiomes (Zettler et al. 2013). Coccolithophores, diatoms, bryozoans, barnacles, dinoflagellates, isopods, cyanobacteria, heterotrophic bacteria, and fungi can all live in plastisphere biofilms (Rogers et al. 2020) (Table 6.2). There is evidence

Table 6.2 Potential microbes associated with the degradation of plastic

Strain	Source	Plastic substrate	References
Serratia marcescens	Ground soil	Film of LLDPE	Azeko et al. (2015)
Achromobacter xylosoxidans	Soil	Film of HDPE	Kowalczyk et al. (2016)
Zalerion maritimum	Marine environment	Pellets of PE	Paço et al. (2017)
Oscillatoria subbrevis *Phormidium lucidum*	Domestic sewage water	Film of LDPE	Sarmah and Rout (2018)
Alcanivorax borkumensis	Mediterranean Sea	Film of LDPE	Delacuvellerie et al. (2019)
Stenotrophomonas panacihumi	Soil of waste storage yard	Film of PP	Jeon and Kim (2016)
Brevibacillus agri *Brevibacillus brevis* *Brevibacillus* sp. *Aneurinibacillus aneurinilyticus*	Landfills and sewage	Film and pellets of PP	Skariyachan et al. (2018)
Rhodococcus sp. *strain* 36 *Bacillus* sp. *strain* 27	Mangrove environments	PP microplastic	Auta et al. (2018)
Acanthopleurobacter pedis *Bacillus cereus* *Pseudomonas otitidis* *Bacillus aerius*	Plastic disposal sites	Film of PVC	Anwar et al. (2016)

(continued)

Table 1 (continued)

Strain	Source	Plastic substrate	References
Phanerocheate chrysosporium	Plastic disposal sites	Film of PVC	Khatoon et al. (2019)
Bacillus sp. AIIW2	Marine environment	Un-plasticized film of PVC	Kumari et al. (2019)
Pseudomonas citronellolis	Soil	Film of plasticized PVC	Giacomucci et al. (2019)

that both MPs and macroplastics are colonized by marine Erythrobacteraceae. Biofouling caused by the buildup of biomass, such as bacteria and phytoplankton, on plastic surfaces usually results in a reduction in hydrophobicity (Rogers et al. 2020). Animals' propensity to confuse MPs for a healthy food source and the particle settling time are both significantly influenced by biofouling. Olfactory and gustatory chemoreceptors in biofilms draw in living things (Kooi et al. 2017). Shorebirds were drawn to the dimethyl sulfide (DMS), which is related to organic molecules, that was formed when biofilm on MP surfaces broke down (Savoca et al. 2016). Clean polystyrene MPs were shown to be consumed by copepods less frequently than unclean or stale polystyrene MPs (Vroom et al. 2017).

6.3 Ecology

Epiplastic biofilms are microbiological communities that are restricted to the plastisphere (Rogers et al. 2020). They have an extensive array of ecological impacts and are completely distinct from the nearby microbial communities (Amaral-Zettler et al. 2020). Their existence may make it easier for MPs to be ingested by aquatic consumers and then spread throughout the food chain (Rummel et al. 2017). This is due to their accumulation in particular places (Botterell et al. 2019), enhancement of olfactory functions (Procter et al. 2019), and other changes in their palatability which increases MP consumption (Goss et al. 2018). Biofilm-induced MP accumulation aided in the trophic transfer (Ward and Kach 2009). MP aggregation made it easier for suspension-feeding bivalves to consume PS, PE, and PP fibers (Porter et al. 2018). In a classic ecological succession, micro- and macro-organisms slowly colonize plastic particle surfaces as biofilms (Bryant et al. 2016). More MPs may be attracted to and trapped in an environmental matrix by aggregates that form as a result of biofilm formation on plastic surfaces (Arias-Andres et al. 2019). The amount of nutrients present in the nearby aquatic habitats affects the microbial hotspots on the plastisphere. It is recognized that a variety of aggregates with anoxic inner cores disintegrate slowly, such as marine snow and fecal pellets (Arias-Andres et al. 2019). A collection produced by interactions between marine bacteria and eukaryotic microorganisms, particularly diatoms, is known as marine snow (Okshevsky et al. 2020). Cyanobacteria, organic matter, and the surrounding seawater are the main factors in the niche partitioning in the plastisphere (Dussud et al. 2018).

6.4 Biogeochemistry

The synthesis and decomposition of organic matter are fundamentally dependent on the biogeochemical cycle. MPs emit dissolved organic carbon (DOC) as a result of abiotic processes that break them down (Romera-Castillo et al. 2018). Their release from MPs through photooxidation or degradation provides a carbon source for the microorganisms (Zhu et al. 2020a, b). As a result, microbial ecology may be influenced by the biogeochemical/nutrient cycling of MPs and the consumption of released DOC by bacteria. The chemistry of the plastic had an impact on how quickly MPs break down or get eliminated. The main byproduct was discovered to be DOC, with EPS degrading the fastest and polypropylene following. Additionally, it was shown that DOC greatly contributed to the ocean's DOC pool and marine food web by encouraging the proliferation of heterotrophic microbes (Rogers et al. 2020). By obstructing light penetration, the CDOM can alter how the marine environment functions and lower its primary output. When compared to MPs buried in sea bottom sand, the biogeochemistry of floating MPs is completely different (Brandon et al. 2019). The main microbial respiration mechanism that governs MP fate is sulfate reduction (Jørgensen and Kasten 2006). Abiotic variables (such as solar radiation and particulate matter) also influence the biogeochemistry of plastic particles in addition to biotic factors (such as microbial biofilms). Large organic particles are produced as a result of PS NPs' interactions with organic materials. According to reports, the settling of very big particles alters the dynamics of aquatic ecosystems. The chemical composition of MPs and NPs may also affect the development of biofilms. Furthermore, amidine-functionalized NPs stopped epiplastic biofilm from developing. Aging MPs can produce a microbiological biofilm and generate chemical metabolic exudates, in contrast to fresh MPs (Vroom et al. 2017). Sand can be used to maintain MPs that are clear of biofilm (He et al. 2020). Proteins, polysaccharides, and humic chemicals found in extracellular polymeric compounds, biofilm-coated sand decreased the Derjaguin-Landau-Verwey-Overbeek (DLVO) aversive contact (He et al. 2020). This makes it clear that the stability of suspended particles is governed by interactions between the electrostatic double layer and Van der Waals forces (Wang et al. 2021). Biofilm increased the hydrophobicity of PE MPs and changed the functional groups that go along with it (Tu et al. 2020). Exposure to plastic particles along with adsorbed chemical pollutants can have a detrimental effect on the anaerobes (hydrogen/methane production) (Chen et al. 2020). However, because PS nano plastic's (NP) harmful effects on microorganisms (i.e., microbe-specific effects) are unpredictable, it is discovered that the suppression of anaerobic digestion is reliant on the concentrations of NP (Fu et al. 2018). FTIR or Raman spectroscopy measures the physical and chemical modifications in plastic polymers during remediation and biodegradation (Wiggin and Holland 2019). It is possible to follow the biodegradation process of plastic polymers using carbon isotopes (Zumstein et al. 2018). Furthermore, organic compounds with the 13C377 designation are crucial to the study of plastic particles (Ledovskaya et al. 2020).

6.5 Genetics

MPs are widely used as substrates by diverse microbial populations and distinct antibiotic resistance genes (ARGs). By analyzing networks and sequencing the 16S rRNA gene, the distinct microbial assemblage was verified. It was discovered that MP-associated bacteria had a higher horizontal transfer frequency of ARGs (Arias-Andres et al. 2018). In this regard, metabolic fingerprinting, next-generation sequencing, qPCR, fluorescent-based flow cytometry, and cell sorting (FACS) techniques are frequently utilized (Pinilla-Redondo et al. 2018). Eventually, EPS-derived microbial cell-plastic aggregates served as hotspots for biogeochemical cycling (Rogers et al. 2020). Horizontal gene transmission (e.g., class 1 integron gene) and the breakdown of organic materials were increased by complexes of MPs and biofilms (Arias-Andres et al. 2019). To understand how MP biofilms' entire resistome of antibiotic-resistant genes affects both animal and human health, more research is required (Arias-Andres et al. 2019). ARGs, islands of pathogenicity, and the metabolic machinery of mobile microbes can all change as a result of plastic-microbe interactions (Rogers et al. 2020). Multiple effects were seen when *Enchytraeus crypticus* was exposed to PS NPs together with tetracycline for 2 weeks (Ma et al. 2020), which were related to molecular (resistome) disruption. Tetracycline ARGs, multidrug ARGs, and macrolide-lincosamide-streptogramin B ARGs, among others, all had a significant impact on *E. crypticus*. Bacterial 16S rRNA gene amplification sequencing showed that this soil invertebrate also had higher levels of aminoglycoside and beta-lactamase ARGs (Ma et al. 2020). The molecular analysis also reported that, even when the exposure was stopped, the abundance and diversity of ARGs could not be fully recovered. Plastisphere biofilms can serve as a source of microbial infections increasing human health risks (Koelmans et al. 2019).

6.6 Ecological Impacts

There are several ecological implications of plastic contaminants in aquatic ecosystems (Fig. 6.2). For instance, they can harm marine phytoplankton by interfering with its photosynthesis, growth, and reproduction (Shen et al. 2020). Due to their negative impacts on marine phytoplankton, which are closely connected to climate change and reduce carbon sequestration and marine production, these factors pose severe dangers to the global ecosystem (Shen et al. 2020). Nanoplastic exposure caused more harmful effects (increased oxidative stress and growth inhibition) in *Halomonas alkaliphila* (Sun et al. 2018). Through feeding chains, zooplankton plays a part in the exchange of genetic information and energy. Particulate organic carbon (POC) and its re-mineralization are eventually impacted by zooplankton, which further affects marine carbon concentrations (Shen et al. 2020). MP contamination interferes with POC mineralization linkages and CO_2 sequestration processes. Therefore, MP contamination eventually affects marine productivity and the capacity for carbon sequestration (Cole 2016). The biogeochemical/nutrient cycle

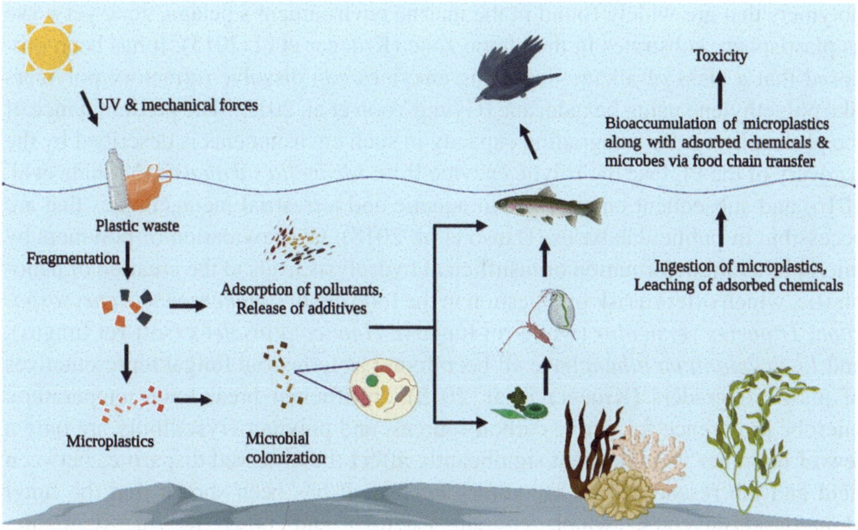

Fig. 6.2 Exposure of plastic to solar UV radiation causes photodegradation combined with physical abrasion and microbial colonization. Aquatic species consume MPs that accumulate via the food chain. With the shrinking size of MPs, their bioaccumulation potential rises. These investigations have sparked concern about their negative impacts on aquatic environments (Chae et al. 2018; Gray and Weinstein 2017; Ziajahromi et al. 2016)

and carbon sequestration processes in aquatic habitats can be impacted by plastic contamination. By monitoring how photosynthetic algae's development is inhibited depending on the concentration (*Scenedesmus obliquus*), it was possible to study the interactions (full factorial) among various stress variables in freshwater (Yang et al. 2020). To clearly show a connection between climate change and nanotoxicology, more research on their interactions is required. The size of the Arctic Sea ice has drastically shrunk during the last 10 years due to climate change (Obbard et al. 2014). Eutrophication (nutrient enrichment in shallow lakes), plastic polymer pollution, and climate change have all been shown to interact (Zhang et al. 2020b). Further explorations should focus on the role of the plastisphere in the modification of polymer density, size, and oxidation state along with the physical and chemical transformation of plastic waste (Dey et al. 2022).

6.7 Bioremediation

Small molecules of low-molecular-weight dissolved organic compounds are released as a result of leaching from plastic, facilitating microbial adherence to the plastic surface and aquatic environment metabolism. The polymers must have a molecular weight that is low enough to pass past the membrane of the microbial cell and undergo intracellular oxidation (Gyung Yoon et al. 2012). Expanded polystyrene, polypropylene, and polyethylene are a few examples of non-hydrolyzable polymers that are widely found in the marine environment's pelagic zone yet serve as plastisphere substrates in the photic zone (Krueger et al. 2015). It has been proposed that a class of alkane-degrading enzymes can dissolve refractory polymers like polyethylene using hexadecane (Gyung Yoon et al. 2012). The predominance of the plastisphere's PET-degrading capacity in such environments is described by the recovery of the PETase hydrolytic enzyme from *Ideonella sakaiensis* (Yoshida et al. 2016) and subsequent enzymes from aquatic and terrestrial metagenomes that are accessible in public databases (Danso et al. 2018). Over-oxidation of polymers by microbial biotransformation or insufficient hydrolysis leads to the creation of nanoplastic, which offers a risk of ingestion in the food chain. *Phanerochaete chrysosporium*, *Trametes versicolor* (white-rot fungus), *Humicola insolens* (soft-rot fungus), and *Enchyodontium album* have all been found as terrestrial fungal representatives of plastic degraders (Krueger et al. 2015). Insufficient breakdown temperature, microbe preference for edible carbon sources, and polymer crystallinity are only a few of the other variables that significantly affect the observed disparities between field and lab research. In a laboratory context, it has been shown that the fungi Homobasidiomycetes, which contain esterase and other polymer-degrading enzymes, may break down nylon and polyethylene (Hibbett and Binder 2001; Shimao 2001). Although they live in mangrove ecosystems, these unusual aquatic species are poorly understood in terms of their metabolism. Any remediation strategy seeks to get rid of the contamination while preserving the environment's natural balance. A helpful method that benefits the environment is phytoremediation (Qi et al. 2018). Recent research results (Li et al. 2019) from a vast number of researchers also corroborated this. Nevertheless, a variety of phytoremediation techniques, including phytofilteration, phytoextraction, and phytostabilization, can be applied to mitigate the pollution. In contaminated soil, live creatures first come into contact with microplastics in the plant roots. When selecting plants for phytoremediation, a variety of characteristics should be taken into account, including MP tolerance, accumulation skill, biomass production, bioaccumulation factor, root system, and others. The main drawback, in addition to the numerous advantages, is the likelihood that microplastics would infiltrate the human food chain. Therefore, take into account whether MP is immobilized or mobilized in the soil while choosing plants (Ebere et al. 2019). Recent studies (Li et al. 2019; Qi et al. 2018) have demonstrated that plants can take MP-related toxins from the soil and act as potential phytoremediators for MP in terrestrial environments. Plant stabilization techniques include the use of sorption and precipitation processes. Groundwater contamination may occur as a result of MP's propensity to infiltrate into deeper soil layers (Fisner et al. 2013). As a result, MP mobility is diminished or even stopped, preventing MP from entering the atmosphere or groundwater, becoming bioavailable, and moving up the food chain. Similar to phytoextraction, the rhizosphere describes the soil that is only a few millimeters thick and directly adjacent to a plant's root surface. The root surface of the plant absorbs or adsorbs contaminants (Fawzy 2008; Gardea-Torresdey et al. 2004). After being soaked, the roots are gathered and properly disposed of. Repeated treatments can bring the site's contamination down to a manageable level (Gardea-Torresdey et al. 2004).

6.8 Conclusion

Therefore, complicated interactions between MP and NP pollution and other anthropogenic environmental stresses are possible. Microbial ecology and the impact of MP-associated biofilms in aquatic habitats can be connected to a variety of ecological disturbances, including modified biogeochemical mechanisms, gaseous hydrocarbons, nutrient cycling, and global concentrations of methane and carbon. It also influences microbial diversity and respiratory processes in aquatic habitats. Interactions between bacteria and MPs inside a plastisphere may potentially result in the biodegradation of the MPs, and the resulting byproducts may enter food webs and subsequently aquatic biogeochemical cycles.

References

Alvim CB, Mendoza-Roca JA, Bes-Piá A (2020) Wastewater treatment plant as microplastics release source–quantification and identification techniques. J Environ Manag 255:109739

Amaral-Zettler LA, Zettler ER, Mincer TJ (2020) Ecology of the plastisphere. Nat Rev Microbiol 18(3):139–151

Anwar MS, Kapri A, Chaudhry V, Mishra A, Ansari MW, Souche Y, Nautiyal CS, Zaidi MGH, Goel R (2016) Response of indigenously developed bacterial consortia in progressive degradation of polyvinyl chloride. Protoplasma 253(4):1023–1032

Arias-Andres M, Klümper U, Rojas-Jimenez K, Grossart HP (2018) Microplastic pollution increases gene exchange in aquatic ecosystems. Environ Pollut 237:253–261

Arias-Andres M, Rojas-Jimenez K, Grossart HP (2019) Collateral effects of microplastic pollution on aquatic microorganisms: an ecological perspective. TrAC Trends Anal Chem 112:234–240

Auta HS, Emenike CU, Jayanthi B, Fauziah SH (2018) Growth kinetics and biodeterioration of polypropylene microplastics by Bacillus sp. and Rhodococcus sp. isolated from mangrove sediment. Mar Pollut Bull 127:15–21

Azeko ST, Etuk-Udo GA, Odusanya OS, Malatesta K, Anuku N, Soboyejo WO (2015) Biodegradation of linear low density polyethylene by Serratia marcescens subsp. marcescens and its cell free extracts. Waste Biomass Valorization 6(6):1047–1057

Ballent A, Corcoran PL, Madden O, Helm PA, Longstaffe FJ (2016) Sources and sinks of microplastics in Canadian Lake Ontario nearshore, tributary and beach sediments. Mar Pollut Bull 110(1):383–395

Botterell ZL, Beaumont N, Dorrington T, Steinke M, Thompson RC, Lindeque PK (2019) Bioavailability and effects of microplastics on marine zooplankton: a review. Environ Pollut 245:98–110

Brandon JA, Jones W, Ohman MD (2019) Multidecadal increase in plastic particles in coastal ocean sediments. Sci Adv 5(9):eaax0587

Bryant JA, Clemente TM, Viviani DA, Fong AA, Thomas KA, Kemp P, Karl DM, White AE, DeLong EF (2016) Diversity and activity of communities inhabiting plastic debris in the North Pacific Gyre. mSystems 1(3):e00024-16

Chae Y, Kim D, Kim SW, An YJ (2018) Trophic transfer and individual impact of nano-sized polystyrene in a four-species freshwater food chain. Sci Rep 8(1):1–11

Chang X, Xue Y, Li J, Zou L, Tang M (2020) Potential health impact of environmental micro-and nanoplastics pollution. J Appl Toxicol 40(1):4–15

Chen W, Yuan D, Shan M, Yang Z, Liu C (2020) Single and combined effects of amino polystyrene and perfluorooctane sulfonate on hydrogen-producing thermophilic bacteria and the interaction mechanisms. Sci Total Environ 703:135015

Cole M (2016) A novel method for preparing microplastic fibers. Sci Rep 6(1):1–7

Danso D, Schmeisser C, Chow J, Zimmermann W, Wei R, Leggewie C, Li X, Hazen T, Streit WR (2018) New insights into the function and global distribution of polyethylene terephthalate (PET)-degrading bacteria and enzymes in marine and terrestrial metagenomes. Appl Environ Microbiol 84(8):e02773-17

de Souza Machado AA, Kloas W, Zarfl C, Hempel S, Rillig MC (2018) Microplastics as an emerging threat to terrestrial ecosystems. Glob Chang Biol 24(4):1405–1416

Dey S, Rout AK, Behera BK, Ghosh K (2022) Plastisphere community assemblage of aquatic environment: plastic-microbe interaction, role in degradation and characterization technologies. Environ Microbiome 17(1):1–21

Dussud C, Meistertzheim AL, Conan P, Pujo-Pay M, George M, Fabre P, Coudane J, Higgs P, Elineau A, Pedrotti ML, Gorsky G (2018) Evidence of niche partitioning among bacteria living on plastics, organic particles and surrounding seawaters. Environ Pollut 236:807–816

Ebere EC, Wirnkor VA, Ngozi VE (2019) Uptake of microplastics by plant: a reason to worry or to be happy? World Sci News 131:256–267

Eriksen M, Mason S, Wilson S, Box C, Zellers A, Edwards W, Farley H, Amato S (2013) Microplastic pollution in the surface waters of the Laurentian Great Lakes. Mar Pollut Bull 77(1–2):177–182

Faure F, Corbaz M, Baecher H, Neuhaus V, de Alencastro L (2013) Pollution due to plastics and microplastics in Lake Geneva. Arch Sci 65:157–164

Fawzy EM (2008) Soil remediation using in situ immobilisation techniques. Chem Ecol 24(2):147–156

Fischer EK, Paglialonga L, Czech E, Tamminga M (2016) Microplastic pollution in lakes and lake shoreline sediments—a case study on Lake Bolsena and Lake Chiusi (central Italy). Environ Pollut 213:648–657

Fisner M, Taniguchi S, Moreira F, Bícego MC, Turra A (2013) Polycyclic aromatic hydrocarbons (PAHs) in plastic pellets: variability in the concentration and composition at different sediment depths in a sandy beach. Mar Pollut Bull 70(1–2):219–226

Fu SF, Ding JN, Zhang Y, Li YF, Zhu R, Yuan XZ, Zou H (2018) Exposure to polystyrene nanoplastic leads to inhibition of anaerobic digestion system. Sci Total Environ 625:64–70

Gardea-Torresdey JL, De La Rosa G, Peralta-Videa JR (2004) Use of phytofiltration technologies in the removal of heavy metals: a review. Pure Appl Chem 76(4):801–813

Giacomucci L, Raddadi N, Soccio M, Lotti N, Fava F (2019) Polyvinyl chloride biodegradation by Pseudomonas citronellolis and Bacillus flexus. New Biotechnol 52:35–41

Gigault J, Ter Halle A, Baudrimont M, Pascal PY, Gauffre F, Phi TL, El Hadri H, Grassl B, Reynaud S (2018) Current opinion: what is a nanoplastic? Environ Pollut 235:1030–1034

Goss H, Jaskiel J, Rotjan R (2018) Thalassia testudinum as a potential vector for incorporating microplastics into benthic marine food webs. Mar Pollut Bull 135:1085–1089

Gray AD, Weinstein JE (2017) Size-and shape-dependent effects of microplastic particles on adult daggerblade grass shrimp (Palaemonetes pugio). Environ Toxicol Chem 36(11):3074–3080

He L, Rong H, Wu D, Li M, Wang C, Tong M (2020) Influence of biofilm on the transport and deposition behaviors of nano- and micro-plastic particles in quartz sand. Water Res 178:115808

Hibbett DS, Binder M (2001) Evolution of marine mushrooms. Biol Bull 201(3):319–322

Huang Y, Liu Q, Jia W, Yan C, Wang J (2020) Agricultural plastic mulching as a source of microplastics in the terrestrial environment. Environ Pollut 260:114096

Inderthal H, Tai SL, Harrison ST (2021) Non-hydrolyzable plastics—an interdisciplinary look at plastic bio-oxidation. Trends Biotechnol 39(1):12–23

Jeon HJ, Kim MN (2016) Isolation of mesophilic bacterium for biodegradation of polypropylene. Int Biodeterior Biodegradation 115:244–249

Jørgensen BB, Kasten S (2006) Sulfur cycling and methane oxidation. In: Marine geochemistry. Springer, Berlin, pp 271–309

Kershaw PJ, Rochman CM (2015) Sources, fate and effects of microplastics in the marine environment: part 2 of a global assessment. Reports and studies-IMO/FAO/Unesco-IOC/WMO/IAEA/UN/UNEP Joint Group of Experts on the Scientific Aspects of Marine Environmental Protection (GESAMP) Eng No. 93

Khatoon N, Jamal A, Ali MI (2019) Lignin peroxidase isoenzyme: a novel approach to biodegrade the toxic synthetic polymer waste. Environ Technol 40(11):1366–1375
Koelmans AA, Nor NHM, Hermsen E, Kooi M, Mintenig SM, De France J (2019) Microplastics in freshwaters and drinking water: critical review and assessment of data quality. Water Res 155:410–422
Kooi M, Nes EHV, Scheffer M, Koelmans AA (2017) Ups and downs in the ocean: effects of biofouling on vertical transport of microplastics. Environ Sci Technol 51(14):7963–7971
Kowalczyk A, Chyc M, Ryszka P, Latowski D (2016) Achromobacter xylosoxidans as a new microorganism strain colonizing high-density polyethylene as a key step to its biodegradation. Environ Sci Pollut Res 23(11):11349–11356
Krueger MC, Harms H, Schlosser D (2015) Prospects for microbiological solutions to environmental pollution with plastics. Appl Microbiol Biotechnol 99(21):8857–8874
Kumari A, Chaudhary DR, Jha B (2019) Destabilization of polyethylene and polyvinylchloride structure by marine bacterial strain. Environ Sci Pollut Res 26(2):1507–1516
Lambert S, Wagner M (2016) Characterisation of nanoplastics during the degradation of polystyrene. Chemosphere 145:265–268
Ledovskaya MS, Voronin VV, Rodygin KS, Ananikov VP (2020) Efficient labeling of organic molecules using 13C elemental carbon: universal access to 13C 2-labeled synthetic building blocks, polymers and pharmaceuticals. Org Chem Front 7(4):638–647
Li L, Zhou Q, Yin N, Tu C, Luo Y (2019) Uptake and accumulation of microplastics in an edible plant. Chin Sci Bull 64(9):928–934
Ma J, Sheng GD, Chen QL, O'Connor P (2020) Do combined nanoscale polystyrene and tetracycline impact on the incidence of resistance genes and microbial community disturbance in Enchytraeus crypticus? J Hazard Mater 387:122012
Mason SA, Kammin L, Eriksen M, Aleid G, Wilson S, Box C, Williamson N, Riley A (2016) Pelagic plastic pollution within the surface waters of Lake Michigan, USA. J Great Lakes Res 42:753–759
Masura J, Baker J, Foster G, Arthur C (2015) Laboratory methods for the analysis of microplastics in the marine environment: recommendations for quantifying synthetic particles in waters and sediments
McCormick A, Hoellein TJ, Mason SA, Schluep J, Kelly JJ (2014) Microplastic is an abundant and distinct microbial habitat in an urban river. Environ Sci Technol 48(20):11863–11871
Nel HA, Dalu T, Wasserman RJ (2018) Sinks and sources: assessing microplastic abundance in river sediment and deposit feeders in an austral temperate urban river system. Sci Total Environ 612:950–956
Ng EL, Lwanga EH, Eldridge SM, Johnston P, Hu HW, Geissen V, Chen D (2018) An overview of microplastic and nanoplastic pollution in agroecosystems. Sci Total Environ 627: 1377–1388
Obbard RW, Sadri S, Wong YQ, Khitun AA, Baker I, Thompson RC (2014) Global warming releases microplastic legacy frozen in Arctic Sea ice. Earths Future 2(6):315–320
Okshevsky M, Gautier E, Farner JM, Schreiber L, Tufenkji N (2020) Biofilm formation by marine bacteria is impacted by concentration and surface functionalization of polystyrene nanoparticles in a species-specific manner. Environ Microbiol Rep 12(2):203–213
Paço A, Duarte K, da Costa JP, Santos PS, Pereira R, Pereira ME, Freitas AC, Duarte AC, Rocha-Santos TA (2017) Biodegradation of polyethylene microplastics by the marine fungus Zalerion maritimum. Sci Total Environ 586:10–15
Pinilla-Redondo R, Cyriaque V, Jacquiod S, Sørensen SJ, Riber L (2018) Monitoring plasmid-mediated horizontal gene transfer in microbiomes: recent advances and future perspectives. Plasmid 99:56–67
Porter A, Lyons BP, Galloway TS, Lewis C (2018) Role of marine snows in microplastic fate and bioavailability. Environ Sci Technol 52(12):7111–7119
Procter J, Hopkins FE, Fileman ES, Lindeque PK (2019) Smells good enough to eat: dimethyl sulfide (DMS) enhances copepod ingestion of microplastics. Mar Pollut Bull 138:1–6

Qi Y, Yang X, Pelaez AM, Lwanga EH, Beriot N, Gertsen H, Garbeva P, Geissen V (2018) Macro- and micro-plastics in soil-plant system: effects of plastic mulch film residues on wheat (Triticum aestivum) growth. Sci Total Environ 645:1048–1056
Ren X, Tang J, Liu X, Liu Q (2020) Effects of microplastics on greenhouse gas emissions and the microbial community in fertilized soil. Environ Pollut 256:113347
Ribeiro F, O'Brien JW, Galloway T, Thomas KV (2019) Accumulation and fate of nano- and micro-plastics and associated contaminants in organisms. TrAC Trends Anal Chem 111:139–147
Rochman CM, Browne MA, Halpern BS, Hentschel BT, Hoh E, Karapanagioti HK, Rios-Mendoza LM, Takada H, Teh S, Thompson RC (2013) Classify plastic waste as hazardous. Nature 494(7436):169–171
Rochman CM, Hentschel BT, Teh SJ (2014) Long-term sorption of metals is similar among plastic types: implications for plastic debris in aquatic environments. PLoS One 9(1):e85433
Rogers KL, Carreres-Calabuig JA, Gorokhova E, Posth NR (2020) Micro-by-micro interactions: how microorganisms influence the fate of marine microplastics. Limnol Oceanogr Lett 5:18–36
Romera-Castillo C, Pinto M, Langer TM, Álvarez-Salgado XA, Herndl GJ (2018) Dissolved organic carbon leaching from plastics stimulates microbial activity in the ocean. Nat Commun 9(1):1–7
Rummel CD, Jahnke A, Gorokhova E, Kühnel D, Schmitt-Jansen M (2017) Impacts of biofilm formation on the fate and potential effects of microplastic in the aquatic environment. Environ Sci Technol Lett 4:258–267
Sarkar DJ, Sarkar SD, Das BK, Manna RK, Behera BK, Samanta S (2019) Spatial distribution of meso and microplastics in the sediments of river Ganga at eastern India. Sci Total Environ 694:133712
Sarmah P, Rout J (2018) Efficient biodegradation of low-density polyethylene by cyanobacteria isolated from submerged polyethylene surface in domestic sewage water. Environ Sci Pollut Res 25(33):33508–33520
Savoca MS, Wohlfeil ME, Ebeler SE, Nevitt GA (2016) Marine plastic debris emits a keystone infochemical for olfactory foraging seabirds. Sci Adv 2(11):e1600395
Shen M, Zhang Y, Zhu Y, Song B, Zeng G, Hu D, Wen X, Ren X (2019) Recent advances in toxicological research of nanoplastics in the environment: a review. Environ Pollut 252:511–521
Shen M, Huang W, Chen M, Song B, Zeng G, Zhang Y (2020) (Micro)plastic crisis: un-ignorable contribution to global greenhouse gas emissions and climate change. J Clean Prod 254:120138
Shimao M (2001) Biodegradation of plastics. Curr Opin Biotechnol 12(3):242–247
Skariyachan S, Patil AA, Shankar A, Manjunath M, Bachappanavar N, Kiran S (2018) Enhanced polymer degradation of polyethylene and polypropylene by novel thermophilic consortia of Brevibacillus sps. and Aneurinibacillus sp. screened from waste management landfills and sewage treatment plants. Polym Degrad Stab 149:52–68
Sruthy S, Ramasamy EV (2017) Microplastic pollution in Vembanad Lake, Kerala, India: the first report of microplastics in lake and estuarine sediments in India. Environ Pollut 222:315–322
Strungaru SA, Jijie R, Nicoara M, Plavan G, Faggio C (2019) Micro-(nano) plastics in freshwater ecosystems: abundance, toxicological impact and quantification methodology. TrAC Trends Anal Chem 110:116–128
Su L, Xue Y, Li L, Yang D, Kolandhasamy P, Li D, Shi H (2016) Microplastics in Taihu lake, China. Environ Pollut 216:711–719
Sun X, Chen B, Li Q, Liu N, Xia B, Zhu L, Qu K (2018) Toxicities of polystyrene nano- and microplastics toward marine bacterium Halomonas alkaliphila. Sci Total Environ 642:1378–1385
Tan X, Yu X, Cai L, Wang J, Peng J (2019) Microplastics and associated PAHs in surface water from the Feilaixia reservoir in the Beijiang River, China. Chemosphere 221:834–840
Tu C, Chen T, Zhou Q, Liu Y, Wei J, Waniek JJ, Luo Y (2020) Biofilm formation and its influences on the properties of microplastics as affected by exposure time and depth in the seawater. Sci Total Environ 734:139237
UNEP (2014) Valuing plastics: the business case for measuring, managing and disclosing plastic use in the consumer goods industry. United Nations Environment Programme, Nairobi

UNEP (2016) Marine plastic debris and microplastics—global lessons and research to inspire action and guide policy change. United Nations Environment Programme, Nairobi

van den Berg P, Huerta-Lwanga E, Corradini F, Geissen V (2020) Sewage sludge application as a vehicle for microplastics in eastern Spanish agricultural soils. Environ Pollut 261:114198

Vroom RJ, Koelmans AA, Besseling E, Halsband C (2017) Aging of microplastics promotes their ingestion by marine zooplankton. Environ Pollut 231:987–996

Wang L, Wu WM, Bolan NS, Tsang DC, Li Y, Qin M, Hou D (2021) Environmental fate, toxicity and risk management strategies of nanoplastics in the environment: current status and future perspectives. J Hazard Mater 401:123415

Ward JE, Kach DJ (2009) Marine aggregates facilitate ingestion of nanoparticles by suspension-feeding bivalves. Mar Environ Res 68(3):137–142

Wiggin KJ, Holland EB (2019) Validation and application of cost and time effective methods for the detection of 3–500 μm sized microplastics in the urban marine and estuarine environments surrounding Long Beach, California. Mar Pollut Bull 143:152–162

Yang Y, Guo Y, O'Brien AM, Lins TF, Rochman CM, Sinton D (2020) Biological responses to climate change and nanoplastics are altered in concert: full-factor screening reveals effects of multiple stressors on primary producers. Environ Sci Technol 54(4):2401–2410

Yoon MG, Jeon HJ, Kim MN (2012) Biodegradation of polyethylene by a soil bacterium and AlkB cloned recombinant cell. J Bioremed Biodegr 3(4):1–8

Yoshida S, Hiraga K, Takehana T, Taniguchi I, Yamaji H, Maeda Y, Toyohara K, Miyamoto K, Kimura Y, Oda K (2016) A bacterium that degrades and assimilates poly(ethylene terephthalate). Science 351(6278):1196–1199

Zettler ER, Mincer TJ, Amaral-Zettler LA (2013) Life in the "plastisphere": microbial communities on plastic marine debris. Environ Sci Technol 47(13):7137–7146

Zhang Y, Kang S, Allen S, Allen D, Gao T, Sillanpää M (2020a) Atmospheric microplastics: a review on current status and perspectives. Earth Sci Rev 203:103118

Zhang Y, Liang J, Zeng G, Tang W, Lu Y, Luo Y, Xing W, Tang N, Ye S, Li X, Huang W (2020b) How climate change and eutrophication interact with microplastic pollution and sediment resuspension in shallow lakes: a review. Sci Total Environ 705:135979

Zhang Y, Lu J, Wu J, Wang J, Luo Y (2020c) Potential risks of microplastics combined with superbugs: enrichment of antibiotic resistant bacteria on the surface of microplastics in mariculture system. Ecotoxicol Environ Saf 187:109852

Zhao S, Zhu L, Wang T, Li D (2014) Suspended microplastics in the surface water of the Yangtze estuary system, China: first observations on occurrence, distribution. Mar Pollut Bull 86(1–2):562–568

Zhu K, Jia H, Sun Y, Dai Y, Zhang C, Guo X, Wang T, Zhu L (2020a) Long-term phototransformation of microplastics under simulated sunlight irradiation in aquatic environments: roles of reactive oxygen species. Water Res 173:115564

Zhu L, Zhao S, Bittar TB, Stubbins A, Li D (2020b) Photochemical dissolution of buoyant microplastics to dissolved organic carbon: rates and microbial impacts. J Hazard Mater 383:121065

Ziajahromi S, Neale PA, Leusch FD (2016) Wastewater treatment plant effluent as a source of microplastics: review of the fate, chemical interactions and potential risks to aquatic organisms. Water Sci Technol 74(10):2253–2269

Zumstein MT, Schintlmeister A, Nelson TF, Baumgartner R, Woebken D, Wagner M, Kohler HPE, McNeill K, Sander M (2018) Biodegradation of synthetic polymers in soils: tracking carbon into CO2 and microbial biomass. Sci Adv 4(7):eaas9024

Biotechnological Tools – Advances and Diagnostics to Control Parasitic Diseases in Aquaculture

7

Shukla Devnath, Souvik Dhar, Anjana Verma,
Vikash Kumar, Suvra Roy, Ajaya Kumar Rout,
Asim Kumar Jana, and Bijay Kumar Behera

Abstract

The aquaculture sector is developing more rapidly, which demands rapid disease diagnostics. The effectiveness of management techniques is significantly influenced by the timely and correct diagnosis of diseases. Devoted farmers and researchers would benefit greatly from adopting quick and on-farm biotechnology techniques for disease diagnosis. The sensitivity of diagnostic tools, epidemiological research, and most crucially the control of these diseases were all greatly improved by molecular techniques. Since the invention of PCR, few significant developments in the field of clinical diagnostic testing have been made, though new technologies are being developed. Molecular technologies like polymerase chain reaction (PCR), real-time PCR, and nucleic acid sequence-based amplification (NASBA) have made it possible to detect, identify, and quantify enormously low levels of water pathogens, and microarray machinery adds a new measurement to multiplex screening for pathogens and response of the host. The use of monoclonal antibodies (mAbs) provides a quick method for identifying infections; antibodies to immunoglobulins from various fish species can be

S. Devnath
Department of Zoology, Vinoba Bhave University Hazaribagh, Hazaribagh, India

Aquatic Environmental Biotechnology and Nanotechnology Division, ICAR-Central Inland Fisheries Research Institute, Barrackpore, Kolkata, West Bengal, India

S. Dhar · S. Roy · A. K. Rout · A. K. Jana · B. K. Behera (✉)
Aquatic Environmental Biotechnology and Nanotechnology Division, ICAR-Central Inland Fisheries Research Institute, Barrackpore, Kolkata, West Bengal, India

A. Verma
Department of Zoology, Vinoba Bhave University Hazaribagh, Hazaribagh, India

V. Kumar
Department of Zoology, Fakir Mohan University, Balasore, Odisha, India

B. K. Behera (ed.), *Current Trends in Fisheries Biotechnology*,
https://doi.org/10.1007/978-981-97-3165-7_7

used to track the host's reaction after immunization, and monoclonal antibodies can screen brood stock for prior pathogen exposure. The goal of this study is to highlight the possibilities of these innovative technologies in the field of diagnosis of parasites, and this study will focus on a new technology that can be used to improve fish health management.

Keywords

Aquaculture · Diagnostics · Biotechnology · Molecular technique · Immunoglobulins · Antibodies

7.1 Introduction

Aquaculture has undergone fast growth, and intensification and disease incidence has risen rapidly along with it. As a result, there is a lot of pressure on researchers and research institutions to concentrate on creating effective diagnostic tools (Jolly et al. 2023). In aquaculture, management is mostly responsible for disease control and prevention, and effective disease control measures heavily rely on accurate and prompt disease detection (Bohara et al. 2024). For disease control in aquaculture to be successful, accurate and prompt illness identification is essential (Subasinghe et al. 2023). In recent years, growing demand for fish along with the opportunity for economic gain has encouraged horizontal and vertical aquaculture expansion worldwide (Gephart et al. 2020). The industry of fisheries provides 17% of animal protein and also provides 12% livelihood of the population of the world (Boyd et al. 2022). India is only second to China in terms of cultured fisheries production with an annual growth rate of over 7%, aquaculture has been increasing quickly in India, where more than 14.5 million people, primarily in coastal areas, depend on it for their daily sustenance (Mamun et al. 2023). According to reports, the aquaculture industry in India has suffered considerable harm due to the frequent incidence of infectious diseases with bacterial and viral origins, impeding sustainable development (Bajpai et al. 2022). Fish's health is affected by a variety of factors such as infections, that are present in aquatic environments (Mishra et al. 2020). The disease becomes apparent when a stressful situation prevails and suppresses the fish's immune system (Bowden et al. 2007).

Numerous applications of biotechnology have been made in the last 10 years in the management of fish health, including the development of fish vaccines as well as diagnostic probes and tests (Mishra et al. 2023). Several laboratory techniques are used in addition to clinical symptoms, host location, and clinical history to detect and diagnose parasite infections. Since the invention of the microscope by Antonie van Leeuwenhoek in the fifteenth century, currently, the primary tests used to identify many parasite infections have not changed. However, recent advancements in diagnostic techniques have created fresh opportunities for a significant improvement in parasite diagnosis (Bohara et al. 2024). The species identification and "strains" (population variants), and infection diagnosis, which aid in the control of parasitism, depend critically on the genetic characterization of parasitic helminths utilizing cutting-edge molecular technologies (Oyesola et al. 2022). Furthermore, investigations of the taxonomy, evolutionary history, and epidemiology of helminths are supported by reliable genetic variation analyses. The choice of appropriate genomic loci is essential to such studies (Thorn et al. 2023).

The techniques of molecular biology are becoming more important in understanding parasitology. Due to the ease and speed with which information about parasite genomes and proteomes is now accessible through numerous databases, including TrypanoCyc, EPICDB LeishCyc, EuPathDB, WormBase, HelmCop GeneDB, PathOD, LeishCyc, and many more such databases that are available on web portals over the past three decades (Kumar et al. 2021). Firstly, extremely sensitive and specific serology-based tests have been developed, including the Dot-ELISA, rapid antigen detection system (RDTS), luciferase immunoprecipitation system (LIPS) and Falcon assay screening test ELISA (FAST-ELISA) (Mukarram et al. 2023).

To improve the sensitivity, and specificity of parasite recognition and the identification of parasite strains, PCR and its variants, including multiplex PCR, nested PCR, LAMP, RAPD, and RACE-PCR, are frequently used (Zhussupova 2022). It can be seen that PCR using TBR and ITS-1 primers can detect *T. evansi* DNA in whole blood at levels as low as 0.1 ng. By quickly screening many parasite genotypes, other approaches like DNA micro-array, DNA probes, and specially designed molecular beacons have a bigger influence on diagnostic and epidemiological research of veterinary parasites (El-Ashram et al. 2019). The availability of a genomic data library is a huge help to parasitologists for gene cloning and sequencing, diagnostics, and the manufacturing of vaccines. Additionally, when compared to different recombinant protein-based vaccines, vaccines like DNA vaccines and subunit vaccines have demonstrated potential for prolonged stimulation of the host immune system.

This study will concentrate on new biotechnology tools that can be used to improve aquatic health and some of the key areas where biotechnology tool has already had a significant impact on fish health management to highlight the potential of these new technology tools in parasite diagnosis.

7.2 Disease Problems in Fish Culture

As natural occupants of the aquatic environment, cultured aquatic animals like fish and shrimp are vulnerable to harmful microorganisms (Mohamad et al. 2019). A complicated relationship between the host species, disease pathogens, and environment results in disease manifestation in aquatic environments. Because of high stocking densities, increased stresses on cultured stocks, and insufficient exchange of water, intensive and semi-intensive agricultural practices encourage the incidence of illness. As a result, fish are more vulnerable to different infectious agents, including viruses, bacteria, helminths, crustaceans, and protozoa (Mugwanya et al. 2022). Sessile ciliates that attach to fish yet consume bacteria, and other nutrients

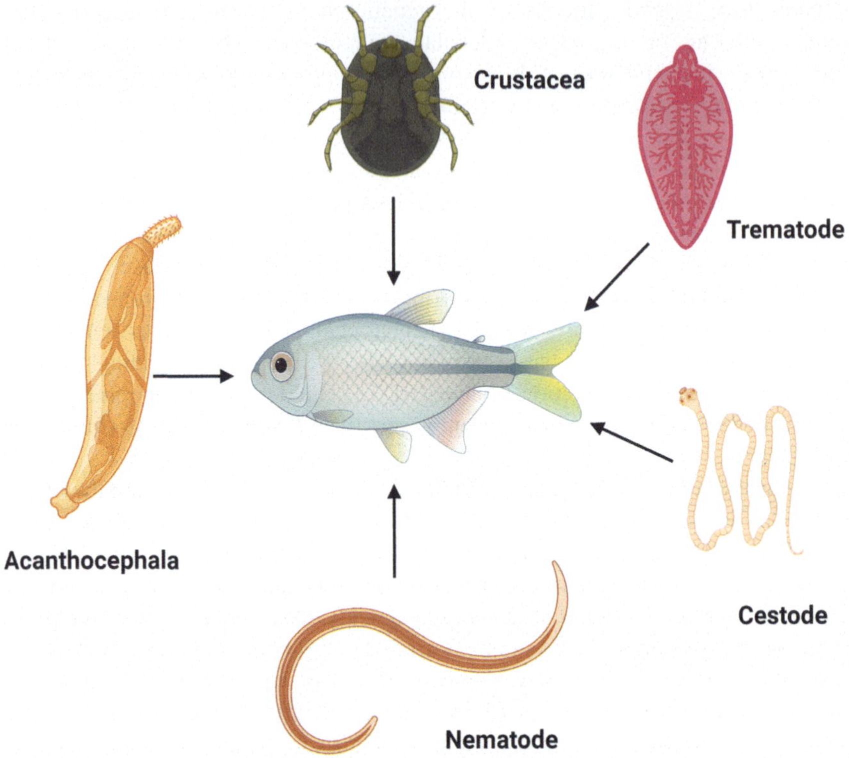

Fig. 7.1 Fish and different metazoan parasites, attack in fish

present in the aquatic environment, such as *Heteropolaria/Epistylis*, *Apiosoma*, and *Ambiphyra*, have been found to thrive best in environments with high levels of organics (Mishra et al. 2020). The trichomonads (such as *Trichodina* and *Tripartiella*), *Tetrahymena,* and *Uronema* protist parasites are other protist parasites that flourish in rich, organic settings (Kostka et al. 2005). In this case *Argulous* infestation is the most common type of parasite among all followed by *Dactylogyrous* species that attack the gills (Paul and Sahoo 2024). Additionally, reports of *Myxobolous*, *Ergasilus*, and *Trichodina* in farmed fish have been made. Numerous bacterial infections also significantly harm fish production (Mishra et al. 2017) (Fig. 7.1).

According to estimates, bacterial pathogens are responsible for about 34% of all infections. These microbes are opportunistic pathogens that attack the tissues of a fish host that has been made vulnerable to infection by stress conditions (Derome et al. 2016). Aquaculture species can become ill from common bacterial infections such as *Vibrio* spp., *Aliivibrio salmonicida, E. tarda*, *Edwardsiella ictaluri*, *Streptococcus* spp., and other similar Gram-positive *Cocci* (Senthamarai et al. 2023) (Table 7.1)

Table 7.1 Some economically important fish parasites, given information on economic loss and mortality caused by fish parasites

Group of parasites	Name of parasites	Disease	Host	Mortality/ economic loss[a]
		Freshwater Endoparasites		
Cnidaria	*T. Bryosalmonae* *M. Cerebralis* *S. Molnari*	Proliferative kidney disease (PKD) Whirling disease Sphaerosporosis	Rainbow trout Rainbow trout Carps	95%[b] 90%[b] NA
Platyhelminthes	*Bothriocephalus acheilognathi*	Ulcer, catarrhal enteritis	Carps	100%[b]
		Freshwater Ectoparasites		
Ciliophora	*I. Multifiliis*	White spot	Rainbow trout, carps	NA
Arthropoda	*Argulus*	Argulosis	Indian major carps	5.41 million USD[a]
Platyhelminthes	*Gyrodactylus salaris*	Gyrodactylosis	Atlantic salmon	NA
		Marine endoparasites		
Cnidaria	*C. Shashta* *E. Leei* *K. Thyrsites*	Ceratomyxosis Enteromyxosis Soft flesh syndrome	Salmonids Sharpsnout seabream Salmonids	NA NA Six million CAD[a]
Ciliophora	*Uronema marinum, Philasterides dicentrarchi*	Scuticociliatosis	Seabass, turbot	NA
		Marine ectoparasites		
Arthropoda	*Caligus rogercresseyi, Lepeophtheirus salmonis*	Sea louse disease	Salmonids	100 million USD[a]
Ciliophora	*C. Irritans*	White spot	Greater amberjack	834 USD[a]

NA not available
[a]Economic loss
[b]Mortality

7.3 Biotechnology Offers a Revolution in Fish Health Management

The quality and quantity of fish produced worldwide are currently significantly impacted by disease losses since culture fish are vulnerable to an extensive range of fungal, viral, parasitic, and bacterial illnesses (Lieke et al. 2020). Rapid pathogen detection and identification, for instance, is essential for efficient illness treatment, which in turn minimizes the antibiotics use and other chemicals in the environment (Vasala et al. 2020). Disease prevention through immunization also dramatically lowers antibiotic consumption. To prevent the spread of disease, it is necessary to restrict the movement of diseased fish, hence regular pathogen detection is crucial (Barnes et al. 2022).

7.4 Diagnostics for Fish Diseases

The quick, accurate, and precise diagnosis of the causal factors of disease outbreaks and mortality forms the basis of a successful health management program (Kumar et al. 2022). The molecular method has considerably developed diagnostic methods not only in humans but also in animal medicine together with fish disease diagnosis and control schemes (Mishra et al. 2020). This method is more efficient than traditional methods as they are typically faster, more specific, and more precise, and previous cultivation of microorganisms is not required (Table 7.2).

7.4.1 Polymerase Chain Reaction

The PCR method was initially developed in 1983 by American biochemist Kary B. Mullis and later developed at the Cetus Corporation in Emeryville, California. PCR is one of the most widely used methods. The influence of PCR on illness diagnosis is unparalleled (Hahn et al. 2020). A thermostable DNA polymerase starts DNA synthesis at a set of two "primers", which designate a specific region of DNA that needs to be amplified. For detection and analysis, a segment of a DNA molecule can be increased one million times (Joshi et al. 2010) (Fig. 7.2). The length of the

Table 7.2 Common molecular tools used in the diagnosis of fish disease and characterization of pathogens

S. no.	Molecular diagnostics for pathogen detection	Genotyping techniques for characterization of pathogens
1	Nested PCR	Amplified fragment length polymorphism (AFLP) assays
2	Polymerase chain reaction (PCR)	Pulse field gel electrophoresis
3	Reverse transcriptase PCR (RT-PCR)	VNT, BOX-PCR, ISPCR, rep-PCR, and ERIC-PCR
4	DNA probe-based hybridization techniques: FISH, in situ hybridization (ISH), fluorescence in situ hybridization, or dot blot, colony hybridization	Amplified ribosomal DNA restriction analysis (ARDRA)
5	Multiplex PCR (mPCR)	Random amplified polymorphic DNA (RAPD) or arbitrarily primed PCR (AP-PCR)
6	Loop-mediated isothermal amplification (LAMP)	DNA sequence analysis
7	Real-time PCR (qPCR)	Ribotyping techniques (16SrDNA)
8	Microarrays	Multilocus sequence typing (MLST) analysis

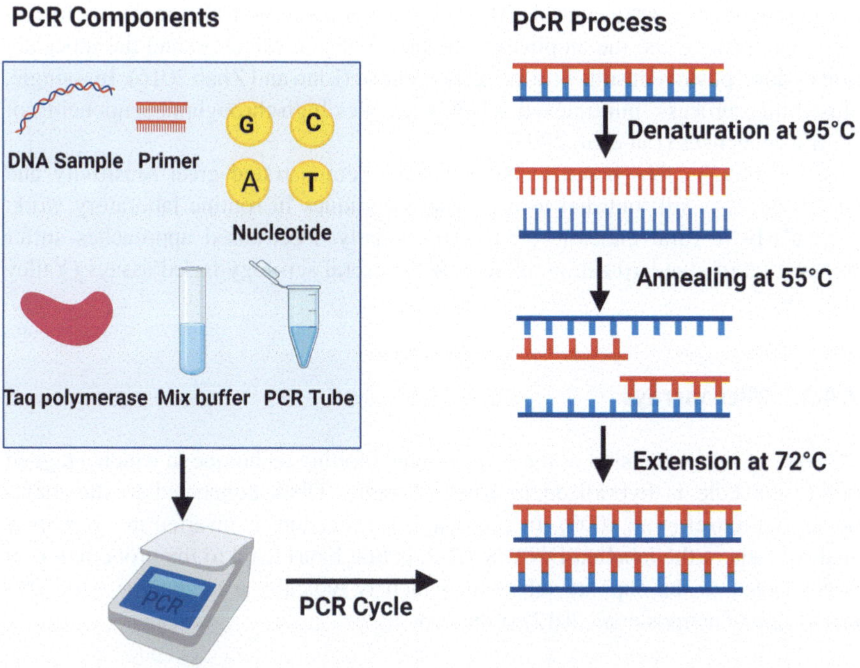

Fig. 7.2 Showing the process of PCR

amplified areas typically falls between 150 and 3000 base pairs (bp). To obtain higher sensitivity and specificity, primer is essential (Rolando et al. 2020). With the introduction of more recent PCR-based approaches, the technology is continuously being improved by many researchers. PCR was widely utilized in the identification of fish viruses including MBV, IHHNV, WSSV, hepatopancreatic parvovirus (HPV), TSV, YHV, etc. since it is incredibly quick, sensitive, and highly specific (Ahamed et al. 2016; Dhar et al. 2019).

7.4.2 Real-Time Polymerase Chain Reaction (RT-PCR)

Further progress in PCR technique is "real-time PCR", also known as qPCR, which allows for real-time monitoring of DNA amplification by the observation of fluorescence at various moments (Li et al. 2020). The most common application of this technology is that it does not require time-consuming post-PCR processing and gel electrophoresis to identify the amplified product (Pal and Pal 2022). Real-time PCR measures the amplicon accumulated by each PCR cycle by measuring the fluorescence produced during the process (Singh and Roy-Chowdhuri 2016).

By utilizing fluorescent chemicals like Taqman probes, Sybergreen, Scorpion primers, and fluorescence resonance energy transfer (FRET), PCR systems, as opposed to conventional PCR, enable the determination of the concentration of the original template (Pestana et al. 2010). By doing away with the necessity for gel electrophoresis to see the amplicons, the possibility of infection and the introduction of false-positive results is drastically reduced (Lian and Zhao 2016). In a single, closed-tube process, multiplexed RT-PCR enables high-throughput sequencing of several sequences (Lai et al. 2007).

Even though DNA-based techniques have demonstrated great sensitivity and specificity, it is still unusual to use these techniques in routine laboratory work, particularly in rural endemic areas. Additionally, PCR-based approaches suffer from a lack of standardization, as seen with several serology-based assays (Yalley et al. 2022).

7.4.3 Microarrays

Microarray is an extension of the conventional blotting technique in which a tagged probe molecule is hybridized to RNA or target DNA connected to the membrane (Schrenzel et al. 2009). In contrast, this procedure is inverted in microarray analysis, where the labelled DNA or RNA is first hybridized to the probe before it is connected to the support substrate, which is typically a nonporous solid surface (López-Campos et al. 2012) (Fig. 7.3).

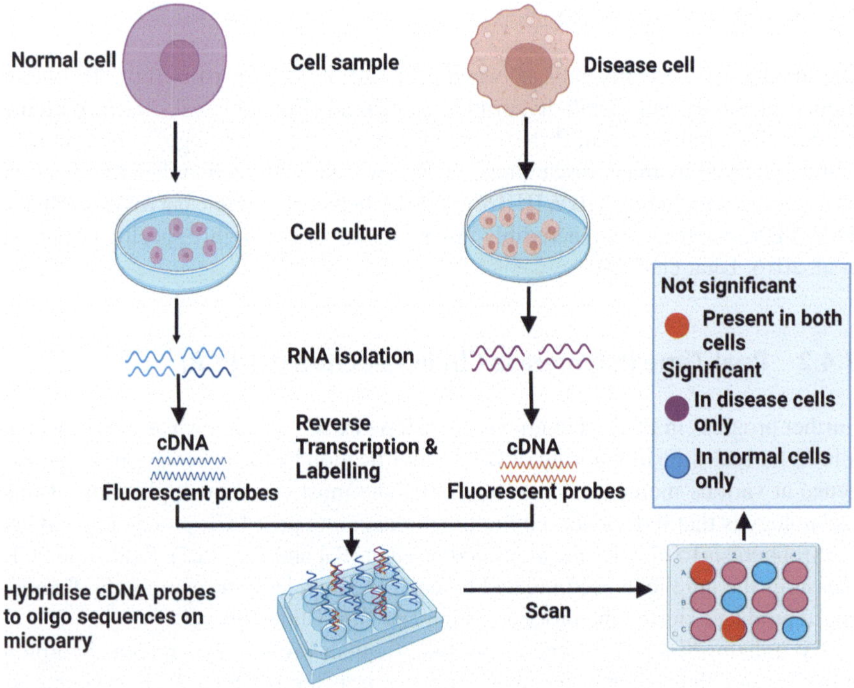

Fig. 7.3 Showing the process of microarray

The material can be simultaneously screened with thousands of probes, which is the main benefit of this method. With this technology, it is possible to evaluate the expression levels of thousands of genes in a given sample all at once. This technology detected 18 microorganisms with 100% specificity, 15 of which were identified as fish pathogens. Similar to this, Cao et al. (2011) recognized a microarray for the instantaneous finding and identification of numerous putative pathogens frequently linked to products of the fishery by focusing specific genes of *Shigella, Listeria monocytogenes, Salmonella, Staphylococcus aureus, Vibrio vulnificus, Vibrio parahaemolyticus, Vibrio cholerae, Streptococcus pyogenes,* and *Yersinia enterocolitica* and the 16S–23S rRNA gene internal transcribed spacer region of *Proteus vulgaris* and *Proteus mirabilis*.

7.4.4 Reverse Transcriptase Polymerase Chain Reaction

Taq polymerase techniques do not function on RNA samples, hence RNA molecules can't be directly amplified by PCR because Taq has low RT activity and DNA polymerase needs an RNA, dsDNA template must be reverse transcribed into cDNA before PCR with the help of the enzyme reverse transcriptase (RT) (Sellner and Turbett 2024).

7.4.5 Luminex xMAP Technology

It is a technology that is based on the bead flow-cytometric assay which enables the recognition of many targets instantaneously. The oligonucleotides, antigens, or antibodies acting as probes in the experiment can be covalently linked to the microsphere beads. The Luminex assay can detect several species or various genotypes of a specific organism in the same reaction with very little volume. Luminex was used in the study of *Cryptosporidium parvum* and *C. hominis* cannot be discriminated against using antigen detection or serology assays. By taking advantage of the single nucleotide difference in the microsatellite-2 region (ML-2) of both species, only DNA-based methods have been successful in doing this (Mukarram et al. 2023). In the end, DNA sequencing is the preferred diagnostic method, although it is expensive, labour and time-intensive.

7.4.6 Nested Polymerase Chain Reaction

It is an alteration of the PCR that aims to decrease non-specific binding with products because of the amplification of unanticipated primer binding sites. Two primer sets are employed in two separate polymerase chain reactions as part of a nested PCR, with the second set being used to amplify a secondary target present in the product of the first run. This limits non-specific products by allowing amplification for a few runs in the initial round. Only the intended product from the first round of amplification should be amplified by the second nested primer set, no other products should be amplified. This enables the execution of more overall cycles while reducing non-specific goods. This is helpful for PCRs with high backgrounds or unusual templates (Green and Sambrook 2019).

7.4.7 Multiplex Polymerase Chain Reaction

Multiplex polymerase chain reaction (Multiplex PCR) is the process of amplifying multiple DNA sequences simultaneously using polymerase chain reaction (as if conducting numerous individual PCR reactions simultaneously in one reaction). Utilizing multiple primers and a thermal cycler with a temperature-mediated DNA polymerase, this procedure amplifies DNA in samples. For all primer pairs to function during PCR at the same annealing temperature, the primer designs must be improved (Markoulatos et al. 2002).

7.4.8 Nanotechnology and Nanosensors

Numerous significant advancements in nearly every branch of science and technology have been made possible by nanoscience and nanotechnology, particularly in biology and medicine (Malik et al. 2023). Due to this technology's small size, nanochips and nanoarrays are now used as test platforms. Current developments in nanotechnology have enabled the creation of simple assays based on nanoparticles for the targeted detection of clinically relevant analytes. The development of quick, precise, and affordable diagnostics for the detection of harmful infectious pathogens is greatly facilitated by nanotechnology (Mishra et al. 2023). Owing to their tiny (1–100 nm) size and huge surface area, nanomaterials exhibit different properties than bulk (micron-size) materials. These variations include, among other things, improved surface reactivity, quantum confinement effects, electrical conductivity, and magnetic characteristics. Biosensors have developed due to the demand for quick identification and characterization of harmful microorganisms. Biosensor performance could be greatly improved by nanotechnology (Poddar and Kishore 2022).

7.4.9 Development of Disease-Resistant Fish Using Marker-Assisted Selection (MAS)

The development of disease-resistant fish strains through marker-assisted selection is one method for preventing fish infections in aquaculture (MAS). Understanding the relationship between DNA markers and the quantitative trait loci (QTL) of a target trait is necessary for MAS. DNA markers can improve a genetic makeup by choosing profitable qualities, such as disease resistance (Yue et al. 2024). Finding

DNA polymorphisms that can be used to track the Mendelian inheritance of homologous chromosomal segments requires using DNA markers (Ma et al. 2024).

7.4.10 Immunological Methods for the Detection of Pathogens

Pathogens can be quickly and precisely detected using immunological techniques such as the western blot, immunohistochemistry (IHC), ELISA, fluorescent antibody technique (FAT), indirect fluorescent antibody technique (IFAT), dot blot, and others (Chauhan et al. 2024).

7.4.11 Antiserum Detection

The use of antibody-based diagnosis appears to be losing significance with the development of easily accessible diagnostic kits and the rapid advancements in genetic manipulation. Animals inoculated with antigens (or immunogens) develop antibodies in reaction to proteins or other molecules the immune system detects as foreign substances (Mishra et al. 2023). Foreign enzymes, carbohydrates, lipids, proteins, viruses, and bacteria make up immunogens. High molecular weight, highly purified proteins identified as foreign bodies and have a high rate of digestion but a low solubility rate make good immunogens.

7.4.12 Agglutination Reaction

Agglutination, easily seen by the clumping of these particles, is an antigen-antibody reaction, particularly between particulate antigens and their specific or cross-reactive antibodies. The quick and simple production of the reaction makes it a very helpful diagnostic tool (Singh 2022).

7.4.13 Immunofluorescence Antibody Test

It is carried out with a fluorescence microscope that uses a separate light source, a mercury lamp, and different wavelengths of light. This light source strikes antibodies linked to fluorescent dyes and allows for visualizing specific colours associated with antigens and antibodies (Cox 2019).

7.4.14 Enzyme-Linked Immunosorbent Assay (ELISA)

The interaction of an antibody with an antigen forms the basis of all immunochemical techniques. Because of its speed, high specificity, and high sensitivity even for small amounts of test samples, the ELISA technique is widely used. There are two

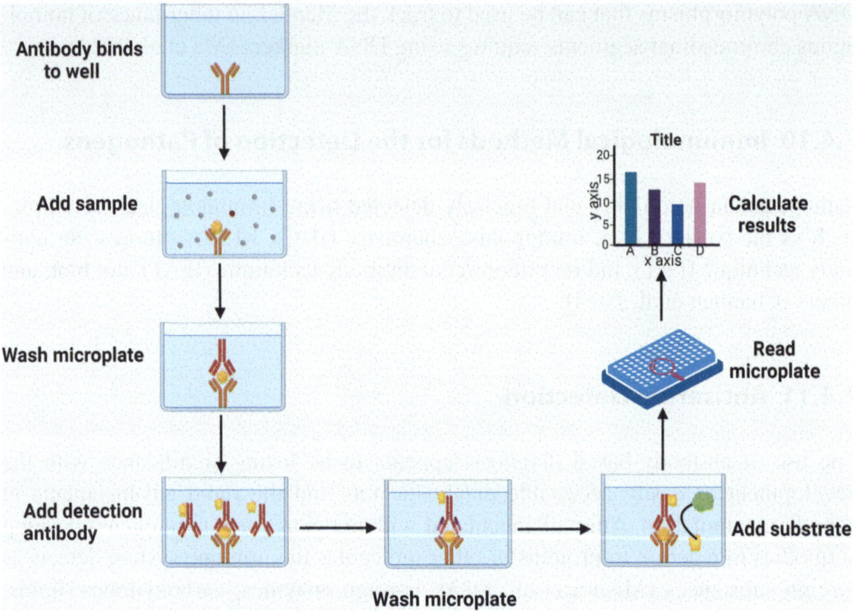

Fig. 7.4 Showing the process of ELISA test

basic methods. The antigen and the antibodies are found using the direct and indirect ELISA methods (Fig. 7.4). A microtiter plate with multiple shallow wells is utilized with the second antibody to develop colour (Polak et al. 2023). Horseradish peroxidase (HRP) is currently the most prevalent, with alkaline phosphatase (ALP) being the oldest. The third enzyme comes from *E. coli* and is called beta-galactosidase. It is active at higher pH levels, which are best for antigen absorption (Ahmed et al. 2023).

7.4.15 Loop-Mediated Isothermal Amplification (LAMP)

Loop-mediated isothermal amplification (LAMP) is an exclusive type of amplification that can distinguish between small differences in a single nucleotide. The LAMP approach has also been used to identify parasite illnesses in fish. *Tetracapsuloides bryosalmonae*, a myxozoan spore, causes proliferative kidney disease (PKD) (Shi et al. 2024). LAMP was used by Et-Matbouli and Soliman to quickly diagnose rainbow trout with (PKD) disease.

7.4.16 Amplified Fragment Length Polymorphism (AFLP)

The PCR-based method known as amplified fragment length polymorphism (AFLP) creates distinct fingerprints for target genomes by selectively amplifying a subset of digested DNA fragments (Nayak et al. 2024). By using these techniques, several parasitic detections were performed.

7.4.17 Random Amplification of Polymorphic DNA (RAPD)

The RAPD method is known as the polymerase chain reaction (PCR) method known as random amplification of polymorphic DNA (RAPD), which is pronounced "rapid" amplifies DNA segments at random. The researcher performing RAPD makes several arbitrary, brief primers (10–12 nucleotides), and then conducts the PCR using a substantial genomic DNA template in the hopes that fragments will amplify. An RAPD reaction can yield a semi-unique profile by resolving the resulting patterns (Bali 2023). An earlier study showed that many parasite identifications were carried out.

7.4.18 Restriction Fragment Length Polymorphism (RFLP)

In molecular biology, restriction fragment length polymorphism (RFLP) is a method that takes advantage of variations in homologous DNA sequences, or polymorphisms, to identify individuals, populations, or species, or to identify the locations of genes within a sequence. The term can refer to a polymorphism itself, as seen in the different locations of restriction enzyme sites, or it can refer to a related laboratory method that can show these variations. For RFLP, the DNA sample is digested by one or more restriction enzymes into fragments analysis, followed by gel electrophoresis to separate the resulting restriction fragments according to their sizes (Ramya and Behera 2023). According to this process, this method is useful against parasitic pathogen detection.

7.4.19 Dot Immunobinding Assay

The dot immunobinding assay is a system of choice for screening on a tight budget since the antigen is affixed to nitrocellulose paper in a series of dots. It is asserted to be either more sensitive or equally sensitive as ELISA assays. The initial procedures required dotting the antigen to nitrocellulose sheets, then cutting up the sheets so that square pieces of paper holding the dot were placed into microtiter wells for blending with the antibody (Cermakova et al. 2023). For identifying and measuring antigens or antibodies, the dot ELISA is a particularly sensitive technique.

7.4.20 Latex Agglutination Test

The very particular interaction of antigen and antibody serves as the foundation for most microsphere-based diagnostic tests (qualitative, yes/no results) and assays.

Polystyrene microspheres that are less than a micron are employed as solid support; antibody or antigen can be adsorbed to them. When antigen in a drop of material reacts with antibody in a drop of uniformly dispersed milky-appearing antibody-coated beads on a glass slide to generate particle agglutination, or the clumping of microspheres, to resemble curdled milk, a positive test ensues (Wang et al. 2022).

7.4.21 Western Blotting

Western blotting is a method for transferring protein from a polyacrylamide gel to a sheet of nitrocellulose in a way that creates an exact copy of the original gel pattern. The SDS-PAGE protein fractionation gel is covered with a sheet of nitrocellulose in this method, and a current is given across the gel to cause the proteins to migrate out of the gel and into the nitrocellulose, where they bind securely by non-covalent forces (Birtwistle et al. 2024). A diagnostic tool for several clinical disorders is western blotting. The western blotting method was utilized to discriminate between different species of piscine trypanosomes.

7.4.22 Monoclonal Antibodies

Monoclonal antibodies are those that are produced by hybridomas. High levels of specificity and infinite production capacity characterize monoclonal antibodies. The three qualities that make monoclonal antibodies helpful are their specificity for binding, homogeneity, and capacity for infinite production. The isolation of agents with a particular, predetermined specificity is made possible by creating monoclonal antibodies. Monoclonal antibodies are effective tools for detecting the presence of the desired epitope since all antibodies produced by a single hybridoma cell are identical (Velázquez et al. 2021). For such studies, monoclonal antibodies provide perfect standardized reagents, and many of them are currently offered commercially. Numerous formats of the ELISA can be used for both serology, the detection of antibodies to the pathogen, and pathogen detection.

7.5 Preventive and Control Measures

7.5.1 Vaccine

Treatment that considers the interactions between the host, environment, and pathogens should take a holistic approach. The major goals of effective management practices are to prevent the disease from entering the farm system, stop the spread of disease agents already present, boost fish natural resistance, and produce healthy, high-quality fish. A recognized, effective, and affordable strategy for controlling several viral illnesses in domesticated animals is vaccination (Gebre et al. 2021).

Vaccines decrease the incidence of disease losses, decrease the need for antibiotic use, leave no product residues, and do not cause pathogen resistance.

Parasite vaccines have been tried, but the aquaculture industry has suffered significantly. Trials of immunization against parasites like *C. multifilis, L. irritans,* and *salmonis*. For protection, many vaccine studies using live, dead, parasite homogenate, subunit, and DNA have been conducted. Multiple fish hosts with multifil infection (Shivam et al. 2021).

The first fish DNA vaccination for industry (against infectious hematopoietic necrosis virus, IHNV), which enables the effective stimulation of cellular responses, is now accessible in Canada. DNA vaccination is a recent key discovery in fish vaccinology. Epitope mapping and proteomics provide an alternate method for identifying vaccine antigens, and the latter offers the possibility of creating peptide vaccines (Mondal and Thomas 2022). Effective fish vaccinations have advanced significantly in recent years. However, there are currently very few commercially viable vaccinations for infectious viral and bacterial illnesses. There are numerous commercially available finfish illness vaccines, and a few more are being developed (Irshath et al. 2023).

Significant bacterial illnesses include those caused by *Aer. salmonicida, Vibrio salmonicida, V. anguillarum, Y. ruckeri, R. salmoninarum, Flavobacterium psychrophilum, F. columnare, P. salmonis, L. garvieae, Streptococcus iniae, Photobacterium damselae* subsp. *Piscicida,* and *E. ictaluri*. Several commercially available vaccinations against viral and bacterial illnesses in fish. *Togaviridae, nodaviridae, orthomyxoviridae*, and other viruses have been the target of more than 20 distinct virus DNA vaccines created experimentally for prophylactic use in fish (Olymon et al. 2024). VHSV and IHNV are two examples of *Rhabdoviridae* DNA vaccines that have demonstrated high levels of efficacy, whereas others have often displayed moderate to low efficacy (Sun et al. 2024). For "Clynav", a DNA vaccine against salmon pancreatic disease (salmon alphavirus-3), the European Medicines Agency (EMA) proposed awarding marketing authorization in the EU in 2016 (López-Porras 2024).

7.5.2 Attenuated Vaccine

A mutant that has lost or diminished its pathogenicity as an antigen is employed with an attenuated vaccination. Attenuated vaccines employ live viruses or bacteria that have just reduced their toxicity; as a result, they are still contagious and may be able to persist in the hosts. Additionally, fewer booster doses are required since attenuated vaccine immunity persists long after inoculation (Andey et al. 2024) (Fig. 7.5).

Other bacterial pathogens with mutant strains have been developed for use as attenuated vaccines, including *Edwardsiella ictaluri*'s crp gene variant, the purA gene mutant strain, the esrB gene mutant strain, and the nutrition-related mutant strain (alr and asd gene mutant) in *E. tarda*; *Flavobacterium psychrophilum* exbD

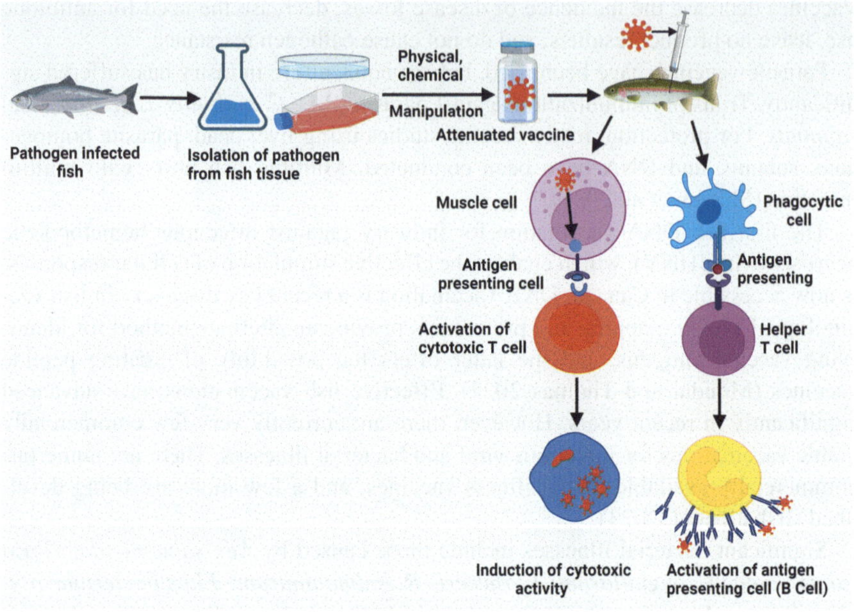

Fig. 7.5 Immune response induced by attenuated vaccine

gene mutant strain Pseudomonas fluorescence variation fur gene mutant strain pgm gene mutant strain and *Streptococcus iniae* simA gene mutant (Aoki 2016).

7.5.3 Subunit Vaccine

Subunit vaccines, also known as component vaccines, are created through genetic engineering using animal cell cultures, yeast, *Bacillus subtilis*, *Escherichia coli*, and yeast, as well as extracted and purified antigenic proteins from pathogenic microbes. It may also be bulk made and is very affordable and safe (Orosco and Espiritu 2024). Unlike the attenuated vaccines previously described, this vaccine's immune response mechanism simply activates the antigen presentation pathway (Fig. 7.6).

7.5.4 DNA Vaccine

In DNA vaccines, recombinant DNA of harmful microorganisms is injected using a gene gun or syringe into the body surface and muscle of fish to induce immunity against pathogenic microorganisms. IHNV DNA vaccinations are now marketed and utilized in Canadian farms However, from the perspective of food safety, since DNA vaccines (recombinant plasmid DNA) are supplied directly to fish and DNA

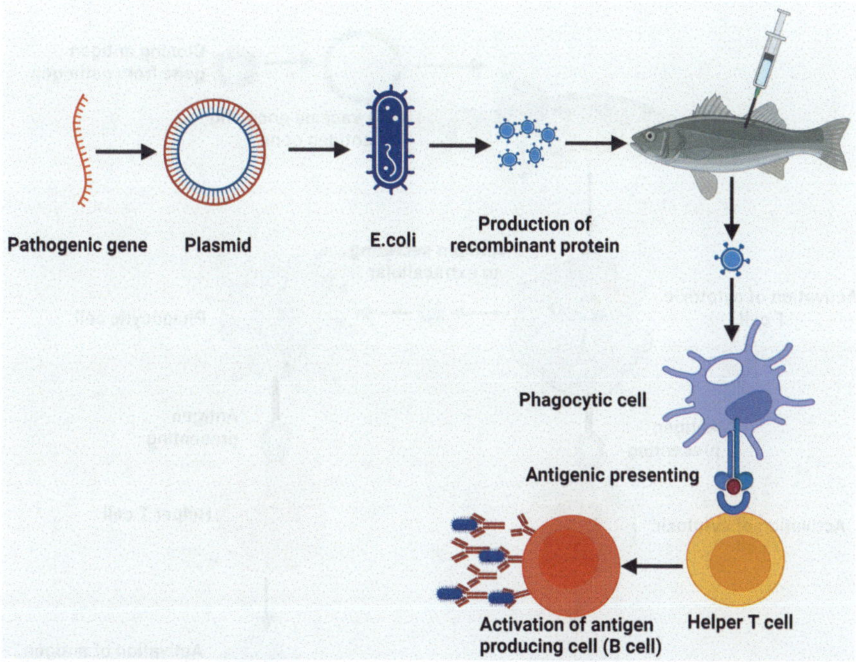

Fig. 7.6 Immune response induced by subunit vaccine

is replicated in the fish body, it is prohibited in all countries except Canada (Naeem et al. 2024).

The effectiveness of DNA vaccines against numerous harmful bacteria and fish parasites has been documented thus far. The effectiveness of the DNA vaccine for the G protein gene against the fish pathogenic viruses IHNV, VHSV, and Hirame Rhabdovirus (HIRRV) has been seen in flounder and salmonid fish (Ju et al. 2024). A systems vaccination strategy utilizing platforms for transcriptomics, epigenetics, proteomics, and metabolomics together with bioinformatics may be required to overcome the difficulty of developing effective vaccinations (Fig. 7.7, Table 7.3).

7.5.5 Fish Immunostimulants

An immunostimulant is a substance that interacts directly with immune system cells and activates them to improve the innate or non-specific immune response. Chemicals, bacterial preparations, polysaccharides, extracts of animals or plants, nutritional components, and cytokines are examples of immunostimulants. The measures of non-specific immunity and resistance to illnesses of fish and crustaceans were found to be improved by immunostimulants (Pothiraj et al. 2024).

Immunostimulants improve innate immune defence processes, not particular immunological responses, to promote resistance to infectious illness. There is no

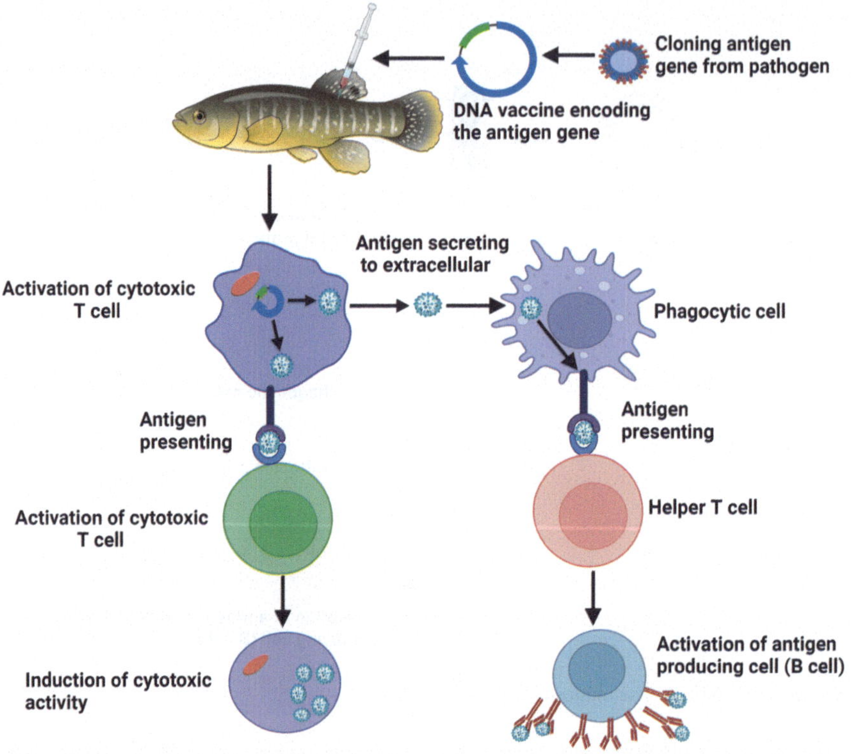

Fig. 7.7 Immune response induced by DNA vaccine

memory component, and the duration of the response is probably brief. In addition to chemotherapy drugs and vaccinations, the use of immunostimulants has been enthusiastically embraced by fish producers. Immunostimulants often stimulate both the innate and acquired immune systems. Immunostimulant-treated fish typically exhibit increased phagocytic cell activities such as phagocytosis, killing, and chemotaxis (Zhang et al. 2024). Immunostimulants also activate NK cells, T cells, and lymphocytes (T and B cells). Immunostimulants can additionally activate humoral components like complement activity and lysozyme (Pothiraj et al. 2024).

Here are a few studies on the interaction between immunostimulants and antibiotics. Tetracycline is one antibiotic that has immunosuppressive properties. To research the resistance to vibrio illness, looked at the interaction between oxytetracycline and glucan. Adjuvants were the initial purpose for the development of immunostimulants. As a result, it can improve antibody production capacity.

Table 7.3 Overview of immunization trials in fish against parasites, given information on immunization trials for endoparasites

Diseases	Parasites	Fish host	Vaccine type	Antigen target
			Freshwater endoparasites	
Myxobolosis	*Myxobolus koi*	Goldfish	Subunit	Crude protein spore
PKD	*Tetracapsuloides bryosalmonae*	Rainbow trout	DNA	Micro-exon gene (TB-MEG1)
			Marine endoparasites	
Scuticociliatosis	*Philasterides dicentrarchi* *Uronema Marinum*	Turbot Grouper	Subunit Subunit	Antigen Membrane protein (i-antigen) immobilization

7.5.6 Probiotics

Probiotics are live, helpful bacteria that, when given in adequate amounts, can improve the host's health (Lian et al. 2024). Probiotics have gained popularity as an antibiotic-free alternative for managing shrimp sickness in aquaculture (Todorov et al. 2024). Probiotics are live microbial feed supplements that benefit the host animal by restoring its microbial equilibrium, according to Fuller (1989). Although probiotics are frequently employed in animal husbandry, their usage in aquaculture is rising due to successful fish and shrimp cultivation in Indian aquaculture.

Nandi et al. (2017) examined how a probiotic-supplemented diet containing *Bacillus amyloliquefaciens CCF7* affected *L. rohita* that was put to the test by a pathogenic strain of *A. hydrophila* and noted changes in the treated group's serum protein, lysozyme, and IgM levels as well as a decrease in stress.

7.5.7 Therapeutics in Fish Diseases

Chemotherapeutics are frequently required to control disease outbreaks, even while improvements in water quality, diet, and other general husbandry aspects may be sufficient to enhance the health of a population (Mishra et al. 2024). Basic practicalities, such as the aquatic environment, fish populations, and delivery routes, as well as the pharmacologic and regulatory difficulties of chemotherapeutic use in fish, make it difficult to manage fish infections (Dinesh et al. 2022). Chemotherapy may make it more difficult to control one's health since it can sometimes negatively affect the environment, public health, and drug resistance (Salam et al. 2023).

7.6 Conclusion

Improvement in the field of rapid diagnostics is essential for the sustainable future of aquaculture. Therefore, a key component of the disease management program is to rapidly and accurately diagnose diseases and infections so that appropriate preventative and control actions can be implemented to control mortality and minimize loss. Traditional diagnostic methods are typically pricy, labour-intensive, and slow and may not result in a conclusive diagnosis of the condition. Additionally, time is lost to conclude so that prompt management action may be implemented. In wild and farmed fish populations, parasitic diseases have recently become more prevalent. Chemotherapy is the standard method for eradicating parasite infestations. Several issues, including the environment's safety, have been brought up in recent years about their use. Vaccines have been suggested by specialists as an efficient way to deal with these problems. Diverse vaccination has demonstrated various levels of parasite protection in fish. Arthropod, protozoal, and helminth genome information provides fresh insights into the biology of parasites and innovative methods for enhancing diagnosis and control strategies. DNA micro-arrays can assist in exploring the new understanding of parasite genomes. Understanding parasite immune evasion mechanisms and host-parasite interactions will help develop successful vaccinations.

References

Ahamed AS, Haq MB, Banu MN, Tiwary C, Sedhuraman R (2016) An update of major DNA shrimp viruses-White Spot Syndrome virus (WSSV), Monodon Baculovius (MBV), Infectious Hypodermal and Hematopoietic Necrosis virus (IHHNV) and Hepatopancreatic Parvovirus (HPV): A Review. Imp J Interdiscip Res 3(1):1–5

Ahmed R, Ray MK, Nayak D, Mohanta YK (2023) Microbial Enzymes in Biomedical Applications. In: InMicrobial Enzymes and Metabolites for Health and Well-Being. CRC Press, pp 53–74

Andey T, Soni S, Modi S (2024) Conventional vaccination methods: Inactivated and live attenuated vaccines. Advanced Vaccination Technologies for Infectious and Chronic Diseases 37-50

Aoki T (2016) editor Fish Diseases. EOLSS Publications

Bajpai V, Pragyan D, Suman K, Mohanty J, Sahoo PK (2022) Viral diseases in Indian freshwater and marine water pisciculture. Curr Sci:267–280

Bali BK (2023) Different Types of DNA Molecular Markers and Genotyping Applications for Mutation Breeding. Biotechnologies and Genetics in Plant Mutation Breeding:239–266

Barnes AC, Rudenko O, Landos M, Dong HT, Lusiastuti A, Phuoc LH, Delamare-Deboutteville J (2022) Autogenous vaccination in aquaculture: a locally enabled solution towards reduction of the global antimicrobial resistance problem. Rev Aquac 14(2):907–918

Birtwistle MR, Huggins JR, Zadeh CO, Sarmah D, Srikanth S, Jones BK, Cascio LN, Dean DR (2024). A 96-Well Polyacrylamide Gel for Electrophoresis and Western Blotting. bioRxiv 2024-05

Bohara K, Joshi P, Acharya KP, Ramena G (2024) Emerging technologies revolutionising disease diagnosis and monitoring in aquatic animal health. Rev Aquac 16(2):836–854

Bowden TJ, Thompson KD, Morgan AL, Gratacap RM, Nikoskelainen S (2007) Seasonal variation and the immune response: a fish perspective. Fish Shellfish Immunol 22(6):695–706

Boyd CE, McNevin AA, Davis RP (2022) The contribution of fisheries and aquaculture to the global protein supply. Food Secur 14(3):805–827

Cao B, Li R, Xiong S, Yao F, Liu X, Wang M, Feng L, Wang L (2011) Use of a DNA microarray for detection and identification of bacterial pathogens associated with fishery products. Appl Environ Microbiol 77(23):8219–8225

Cermakova E, Lencova S, Mukherjee S, Horka P, Vobruba S, Demnerova K, Zdenkova K (2023) Identification of fish species and targeted genetic modifications based on DNA analysis: State of the art. Food Secur 12(1):228

Chauhan RS, Malik YS, Saminathan M, Tripathi BN (2024) Immunopathology of Reproductive Disorders of Animals. In: InEssentials of Veterinary Immunology and Immunopathology. Springer Nature Singapore, Singapore, pp 293–372

Cox G (2019) editor. CRC Press, Fundamentals of fluorescence imaging

Derome N, Gauthier J, Boutin S, Llewellyn M (2016) Bacterial opportunistic pathogens of fish. When commensals and symbionts become parasitic, The Rasputin effect, pp 81–108

Dhar AK, Cruz-Flores R, Caro LF, Siewiora HM, Jory D (2019) Diversity of single-stranded DNA containing viruses in shrimp. VirusDisease 30:43–57

Dinesh R, Anand C, John KR, George MR, Bharathi S, Kumar JS (2022) An overview of chemicals and drugs in aquaculture disease management. Indian Journal of Animal Health 62(1):1–20

El-Ashram SA, Amin AS, Behour TS, Mousa WM, Aboelhadid SM (2019) Molecular diagnosis of acute and chronic infection of Trypanosoma evansi in experimental male and female mice. Onderstepoort J Vet Res 86(1):1–0

Gebre MS, Brito LA, Tostanoski LH, Edwards DK, Carfi A, Barouch DH (2021) Novel approaches for vaccine development. Cell 184(6):1589–1603

Gephart JA, Golden CD, Asche F, Belton B, Brugere C, Froehlich HE, Fry JP, Halpern BS, Hicks CC, Jones RC, Klinger DH (2020) Scenarios for global aquaculture and its role in human nutrition. Rev Fish Sci Aquac 29(1):122–138

Green MR, Sambrook J (2019) Nested polymerase chain reaction (PCR). Cold Spring Harb Protoc 2019(2):pdb-rot095182

Hahn A, Podbielski A, Meyer T, Zautner AE, Loderstädt U, Schwarz NG, Krüger A, Cadar D, Frickmann H (2020) On detection thresholds–a review on diagnostic approaches in the infectious disease laboratory and the interpretation of their results. Acta Trop 205:105377

Irshath AA, Rajan AP, Vimal S, Prabhakaran VS, Ganesan R (2023) Bacterial pathogenesis in various fish diseases: recent advances and specific challenges in vaccine development. Vaccines 11(2):470

Jolly CM, Nyandat B, Yang Z, Ridler N, Matias F, Zhang Z, Murekezi P, Menezes A (2023) Dynamics of aquaculture governance. J World Aquacult Soc 54(2):427–481

Joshi M, Deshpande JD (2010) Polymerase chain reaction: methods, principles and application. Int J Biomed Res 2(1):81–97

Ju PM, Ma SW, Li YY, Zhang SF, Li J, Zhou GZ (2024) Investigation of the Antiviral Mechanism of Curcumin Analog EF-24 against Siniperca cachuatsi Rhabdovirus. Aust Fish 9(5):179

Kostka M, Hampl V, Cepicka I, Flegr J (2005) Parasitic Stramenopila. J Eukaryot Microbiol 52(2):35S–38S

Kumar S, Gupta S, Mohmad A, Fular A, Parthasarathi BC, Chaubey AK (2021) Molecular tools-advances, opportunities and prospects for the control of parasites of veterinary importance. Int J Trop Insect Sci 41:33–42

Kumar V, Roy S, Behera BK, Das BK (2022) Disease diagnostic tools for health management in aquaculture. In: Advances in Fisheries Biotechnology. Singapore, Springer Nature Singapore, pp 363–382

Lai YL, Chung YK, Tan HC, Yap HF, Yap G, Ooi EE, Ng LC (2007) Cost-effective real-time reverse transcriptase PCR (RT-PCR) to screen for Dengue virus followed by rapid single-tube multiplex RT-PCR for serotyping of the virus. J Clin Microbiol 45(3):935–941

Li X, Nguyen LV, Hill K, Ebendorff-Heidepriem H, Schartner EP, Zhao Y, Zhou X, Zhang Y, Warren-Smith SC (2020) All-fiber all-optical quantitative polymerase chain reaction (qPCR). Sensor Actuat B-Chem 323:128681

Lian DS, Zhao SJ (2016) Capillary electrophoresis based on nucleic acid detection for diagnosing human infectious disease. Clinical Chemistry and Laboratory Medicine (CCLM) 54(5):707–738

Lian X, Shi M, Liang Y, Lin Q, Zhang L (2024) The Effects of Unconventional Feed Fermentation on Intestinal Oxidative Stress in Animals. Antioxidants 13(3):305

Lieke T, Meinelt T, Hoseinifar SH, Pan B, Straus DL, Steinberg CE (2020) Sustainable aquaculture requires environmental-friendly treatment strategies for fish diseases. Rev Aquac 12(2):943–965

López-Campos G, Martínez-Suárez JV, Aguado-Urda M, López-Alonso V, López-Campos G, Martínez-Suárez JV, Aguado-Urda M, López-Alonso V (2012) DNA microarrays: Principles and technologies. Microarray Detection and Characterization of Bacterial Foodborne Pathogens:33–47

López-Porras A (2024) Adaptive Humoral Immunity in Atlantic cod (Gadus morhua), a Teleost Fish with Evolutionary Losses of MHC II and CD4

Ma D, Lin Y, Zhang R, Wang S, Hu W, Ye M, Gao H, Wang L, Song Y, Guo H (2024) Effect of uniparental disomy in parentage testing. Legal Med 67:102381

Malik S, Muhammad K, Waheed Y (2023) Nanotechnology: A revolution in modern industry. Molecules 28(2):661

Mamun AA, Krishnan M, Pandian K, Parappurathu S, Parappurathu S, Menon M, Jeeva C, Belevendran J, Anirudhan A, Lekshmi PS, Ramachandran C (2023) Sustainable intensification of small-scale mariculture systems: Farm-level insights from the coastal regions of India. Socio-economic evaluation of cropping systems for smallholder farmers–challenges and options p 88

Markoulatos P, Siafakas N, Moncany M (2002) Multiplex polymerase chain reaction: a practical approach. J Clin Lab Anal 16(1):47–51

Mishra A, Kim HS, Kumar R, Srivastava V (2024) Advances in Vibrio-related infection management: an integrated technology approach for aquaculture and human health. Crit Rev Biotechnol:1–28

Mishra S, Seshagiri B, Rathod R, Sahoo SN, Choudhary P, Patel S, Behera DK, Ojha DK, Jena A, Namburu PK, Swain P (2023) Recent advances in fish disease diagnosis, therapeutics, and vaccine development. Front Aquac. Biotech:115–145

Mishra SS, Das R, Sahoo SN, Swain P (2020) Biotechnological tools in diagnosis and control of emerging fish and shellfish diseases. In: InGenomics and biotechnological advances in veterinary, poultry, and fisheries. Academic Press, pp 311–360

Mishra SS, Rakesh D, Dhiman M, Choudhary P, Debbarma J, Sahoo SN, Mishra CK (2017) Present status of fish disease management in freshwater aquaculture in India: state-of-the-art-review. J Aquac Fish 1(003):14

Mohamad N, Amal MN, Yasin IS, Saad MZ, Nasruddin NS, Al-saari N, Mino S, Sawabe T (2019) Vibriosis in cultured marine fishes: a review. Aquaculture 512:734289

Mondal H, Thomas J (2022) A review on the recent advances and application of vaccines against fish pathogens in aquaculture. Aquac Int 30(4):1971–2000

Mugwanya M, Dawood MA, Kimera F, Sewilam H (2022) A review on recirculating aquaculture system: Influence of stocking density on fish and crustacean behavior, growth performance, and immunity. Ann Anim Sci 22(3):873–884

Mukarram Shah SM, Saira HF (2023) Molecular Techniques for the Study and Diagnosis of Parasite Infection. Immune Responses and Therapeutics, Parasitic Infections, pp 176–204

Naeem S, Hameed R, Sadaqat S, Shahid N, Latif A, Yasmeen A, Azam S, Rao AQ (2024) Microalgae as a Bioreactor for Molecular Farming for the Production of Oral Edible Vaccines Against Infectious Diseases of Humans and Animals. In: InApplications of Plant Molecular Farming. Springer Nature Singapore, Singapore, pp 631–647

Nandi A, Banerjee G, Dan SK, Ghosh K, Ray AK (2017) Probiotic efficiency of *Bacillus* sp. in *Labeo rohita* challenged by *Aeromonas hydrophila*: assessment of stress profile, haemato-biochemical parameters and immune responses. Aquac Res 48(8):4334–4345

Nayak G, Sibadatta A, Suvadra JS, Dash M (2024) Insights into Molecular Markers and Applications in the 21st Century. In: InSmart Breeding. Apple Academic Press, pp 1–30
Olymon K, Yadav M, Teronpi V, Kumar A (2024) Unravelling the genomic secrets of bacterial fish pathogens: a roadmap to aquaculture sustainability. Mol Biol Rep 51(1):364
Orosco FL, Espiritu LM (2024) Navigating the landscape of adjuvants for subunit vaccines: recent advances and future perspectives. Int. J. Appl, Pharmaceut, p 10
Oyesola OO, Souza CO, Loke PN (2022) The influence of genetic and environmental factors and their interactions on immune response to helminth infections. Front Immunol 13:869163
Pal A, Pal A (2022) Real-time or quantitative PCR. Protocols in Advanced Genomics and Allied Techniques181-209
Paul A, Sahoo PK (2024) A host choosy gill parasite (Dactylogyrus spp.) in fish: an insight into host-parasite interaction for developing control strategies. Aquac Int 1-27.
Pestana E, Belak S, Diallo A, Crowther JR, Viljoen GJ, Pestana EA, Belak S, Diallo A, Crowther JR, Viljoen GJ (2010) Real-time PCR–the basic principles. Early, rapid and sensitive veterinary molecular diagnostics-real time PCR applications:27–46
Poddar K, Kishore AV (2022) Nanotechnology in animal production. In: InEmerging Issues in Climate Smart Livestock Production. Academic Press, pp 149–170
Polak JM, Van Noorden S, Polak DJ, Van Noorden S (2023) Introduction to immunocytochemistry. Garland Science
Pothiraj C, Jyoti D, Ramya S, Jayakumararaj R, Grover A, Sinha R, Ayyappan P, Faggio C, Balaji P (2024) Disease Management and Prophylaxis by Immunostimulants. In: InImmunomodulators in Aquaculture and Fish Health. CRC Press, pp 89–102
Ramya VL, Behera BK (2023) Molecular Markers and Their Application in Fisheries and Aquaculture. In: InBiotechnological Tools in Fisheries and Aquatic Health Management. Springer Nature Singapore, Singapore, pp 115–150
Rolando JC, Jue E, Barlow JT, Ismagilov RF (2020) Real-time kinetics and high-resolution melt curves in single-molecule digital LAMP to differentiate and study specific and non-specific amplification. Nucleic Acids Res 48(7):e42–e42
Salam MA, Al-Amin MY, Salam MT, Pawar JS, Akhter N, Rabaan AA, Alqumber MA (2023) Antimicrobial resistance: a growing serious threat for global public health. InHealthcare (Vol. 11, No. 13, p. 1946) MDPI
Schrenzel J, Kostic T, Bodrossy L, Francois P (2009) Introduction to microarray-based detection methods. Detection of Highly Dangerous Pathogens: Microarray Methods for BSL 3:1–34
Sellner LN, Turbett GR (2024) Comparison of three RT-PCR methods. BioTechniques 25(2)
Senthamarai MD, Rajan MR, Bharathi PV (2023) Current risks of microbial infections in fish and their prevention methods: A review. Microb Pathog 185:106400
Shi M, Cao Y, Hu X, Luo F, Lv A (2024) Development of RT-qPCR and loop-mediated isothermal amplification (LAMP) assays for the rapid detection of Rahnella aquatilis in fish. Aquaculture 578:740117
Shivam S, El-Matbouli M, Kumar G (2021) Development of fish parasite vaccines in the OMICs era: Progress and opportunities. Vaccine 9(2):179
Singh C, Roy-Chowdhuri S (2016) Quantitative real-time PCR: recent advances. Clinical Applications of PCR:161–176
Singh J (2022) Antigen–Antibody Interactions. InAn Interplay of Cellular and Molecular Components of Immunology (pp.133-149) CRC Press
Subasinghe R, Alday-Sanz V, Bondad-Reantaso MG, Jie H, Shinn AP, Sorgeloos P (2023) Biosecurity: Reducing the burden of disease. J World Aquacult Soc 54(2):397–426
Sun H, Huang J, Wang H, Zhang Y, Fei Q, Zhou J, Yang L, Li Y, Cai S, Huang Y (2024) Mass mortality associated with Carpione Rhabdovirus in golden pompano (Trachinotus ovatus) in China: First report. bioRxiv 2024-05
Thorn CS, Maness RW, Hulke JM, Delmore KE, Criscione CD (2023) Population genomics of helminth parasites. J Helminthol 97:e29
Todorov SD, Lima JM, Bucheli JE, Popov IV, Tiwari SK, Chikindas ML (2024) Probiotics for Aquaculture: Hope, Truth, and Reality. Probiotics and Antimicrobial Proteins:1–4

Vasala A, Hytönen VP, Laitinen OH (2020) Modern tools for rapid diagnostics of antimicrobial resistance. Front Cell Infect Microbiol 10:308

Velázquez J, Rodríguez A, Aragón H, Haidar A, González M, Valdés R, Garay HE, Abreu DD, Ramos Y, Cabrales A, Morales A (2021) Monoclonal antibody against Nile tilapia (Oreochromis niloticus) IgM heavy chain: A valuable tool for detection and quantification of IgM and IgM+ cells. Fish Shellfish Immunol 110:44–54

Wang Z, Zhao J, Xu X, Guo L, Xu L, Sun M, Hu S, Kuang H, Xu C, Li A (2022) An overview for the nanoparticles-based quantitative lateral flow assay. Small Methods 6(1):2101143

Yalley AK, Ahiatrogah S, Kafintu-Kwashie AA, Amegatcher G, Prah D, Botwe AK, Adusei-Poku MA, Obodai E, Nii-Trebi NI (2022) A systematic review on suitability of molecular techniques for diagnosis and research into infectious diseases of concern in resource-limited settings. Current Issues in Molecular Biology 44(10):4367–4385

Yue GH, Wang L, Sun F, Yang ZT, Wong J, Wen YF, Pang HY, Lee M, Yeo ST, Liang B, Chen K (2024) Improving growth, omega-3 contents, and disease resistance of Asian seabass: status of a 20-year family-based breeding program. Rev Fish Biol Fish 34(1):91–110

Zhang Q, Yang H, Teame T, Ran C, Yang Y, Yao Y, Ding Q, Liu S, Li S, Zhang Z, Zhou Z (2024) Immune disorders induced by improper use of dietary immunostimulants in aquatic animals: Research progress and prospective solutions by targeting gut microbiota. Rev Aquac 16(2):608–621

Zhussupova A (2022) PCR–diagnostics. Litres

Environmental DNA: A Potential Tool in Biodiversity Conservation

8

Prajna Ritambhara Swain, Pritijyoti Majhi,
Pranaya Kumar Parida, and Partha Sarathi Swain

The truth is: the natural world is changing. And we are totally dependent on that world. It provides our food, water and air. It is the most precious thing we have and we need to defend it.

Sir David Attenborough

Abstract

Environment DNA or eDNA studies are regarded as one of the most advanced and powerful tools in the current era to study biodiversity. Not only this technique is useful to study on current status of biodiversity, but also it is well adapted to study from the retro perspective. In biodiversity-related aspects, eDNA has been used to study biomass, species composition, detection of invasive species, identification of unknown species, monitoring the rare and threatened species, study of ancient environments and extinction, etc. Although eDNA is advanced over the traditional methodologies, it has certain drawbacks or limitations. So, judicious uses are required while using these types of techniques. In this chapter, we have discussed about certain uses and common drawbacks of e-DNA and also emphasized some newer additions to this technology to make this technology even more useful.

P. R. Swain (✉) · P. Majhi · P. K. Parida
ICAR-Central Inland Fisheries Research Institute, Barrackpore, Kolkata, India

P. S. Swain
Assistant Fisheries Officer, Office of Additional Fisheries Officer (Marine), Balasore, Odisha, India

B. K. Behera (ed.), *Current Trends in Fisheries Biotechnology*,
https://doi.org/10.1007/978-981-97-3165-7_8

Keywords

eDNA · Biodiversity · Sed DNA · DNA decay · Biomass estimation

It's true, that we are part of the changing world; from birth to the last breath nature nurtures us through its countless biological diversity and ecological services. Human civilization has developed many folds from the earlier *Australopithecus afarensis* (nicknamed Lucy) to the current *Homo sapiens*. The journey of biodiversity has not started with lucy, even from the Hadean eon-based LUCA (last universal common ancestor) to the current Phanerozoic eon (humans), the earth has changed a lot and many of them have also extinct from the earth's surface. In the current Anthropocene time period, the extinction rate has increased surpassing the pre-human period and it is continuously increasing (Dirzo et al. 2022). It not only affects the sustainability of the earth but also impacts human health (Díaz et al. 2006). Despite many international programs and agreements like the earth Summit, Man and Biosphere (MAB) and United Nation Environment Program (UNEP) the loss has not stopped to a greater extent. All such programs more or less depend on the information of biodiversity across the world, including many undefined numerous taxa of varied geographical regions of earth and information on conservation efforts like species monitoring, distribution of species as well as the population across the time frame (sometimes it is very difficult to assess this type of information). Although the species monitoring has been done by using the traditional visual identification by measuring and counting the morphomeristic characteristics; this type of technology sometimes shows ambiguity about species identification during some situations like if the species are closely related in appearance during different life stages and species showing phenotypic plasticity. Eventually, it will create species databases flawed with errors (Daan 2001). Sometimes, the traditional technologies are found to be invasive, e.g., surveys in the marine sector (Baldwin et al. 1996) and the traditional technology also requires skilled taxonomy expert, which is in decline rate (Hopkins and Freckleton 2002; Wheeler et al. 2004). So, ecological monitoring or surveys require an alternative tech-savvy improved monitoring approach.

A contemporary technique like eDNA or environmental DNA eases the process of collecting information on species, populations, and communities of any ecosystem from the analyzed DNA information, which is collected from the environment samples (Baird and Hajibabaei 2012; Kelly et al. 2014); overcoming all the mentioned obstacles. An illustration of eDNA-based species detection is depicted in Fig. 8.1. As we know, organisms living in a habitat expel their DNA to their background/habitat due to their continuous interaction with the ecosystem and the leaching of dead material during decomposition. The DNA can enter the habitat through excreted cells or tissues or by means of urine, feces, hair, and skin (Poinar et al. 1998; Valiere and Taberlet 2000; Bunce et al. 2005; Lydolph et al. 2005). The microbial eDNA is generally found inside the mitochondria (Turner et al. 2014), but its presence in the environment happened only due to membrane degradation (Nielsen

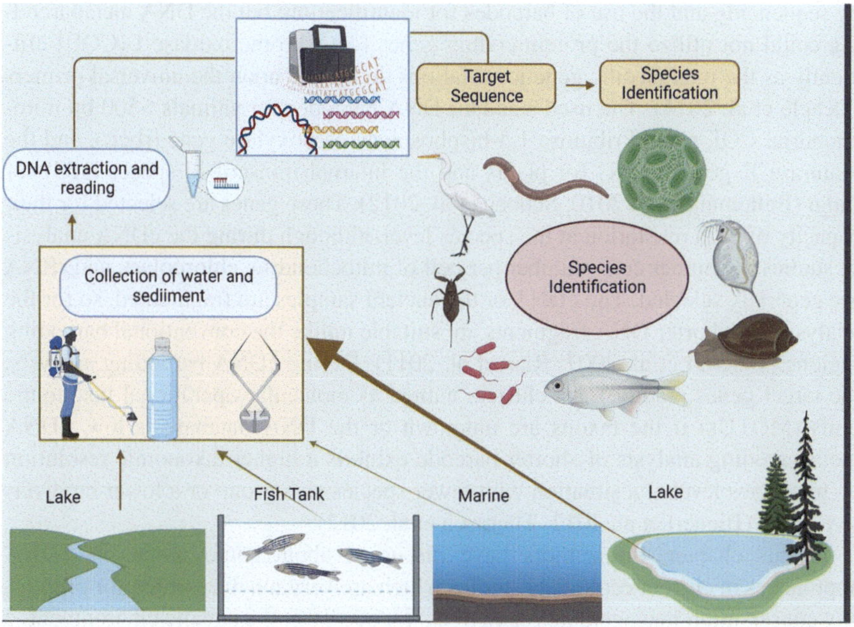

Fig. 8.1 Schematic diagram indicating various processes involved in eDNA analysis such as the collection of sediment and water sample from various aquatic environments, eDNA extraction, and the target sequencing process for identification of species in a particular environment

et al. 2007). The presence of eDNA in the environment varied from some days in water (Thomsen et al. 2012) to thousands of years in cold, permafrost, or polar regions (Willerslev et al. 2003).

EDNA offers a myriad of applications in diverse fields such as molecular biology, ecology, conservation biology, paleontology, and many aspects of environmental sciences. The tiny single standard DNA contains whole information about the ecosystem across the inhabiting communities. For getting the information, the process begins with amplification of the DNA sample through the PCR or polymerase chain reaction, followed by sequencing. The process of amplification might be carried out by single species approach using the species-specific primer or multi-species or multi-taxon approach using the available generic primer. Nowadays researchers are using the most advanced next-generation sequence (NGS) technology, which is a comparatively easy and cost-effective technique (Shokralla et al. 2012). The NGS is quite helpful for studying the community ecology through DNA metabarcoding or mass DNA sequencing, which simultaneously identifies multi taxa from a complex sample (Taberlet et al. 2012).

The process of e-DNA analysis also follows the same principle as the traditional DNA barcoding method of DNA extractions (Hebert et al. 2003), whereas it differs from the later techniques as per its utility and target sequence concerned. In both of the techniques, the short-standardized DNA (whether it is mitochondria, chloroplast, or ribosomal RNA) undergoes the process of amplification by PCR followed by sequencing and the use of barcodes for identification; but the DNA metabarcoding could not utilize the protein-coding genes [cytochrome oxidase I (COI)] efficiently as the interspecific genetic variations do not act upon the universal primers (Deagle et al. 2014). The used standard DNA barcodes for animals >500 bp mitochondrial COI, plastid ribulose 1,5-bisphosphate carboxylase gene (rbcL), and the maturase K gene (matK) for plants and the internal transcribed spacer (ITS) for fungi (Bellemain et al. 2010; Schoch et al. 2012). These genes are selected for their capacity of high resolution at the species level, although during the eDNA analysis or studies the higher copy number per cell of mitochondria, chloroplast, and rRNA are generally selected. The eDNA or the ancient samples are fragmented, so for the analysis, the shorter DNA fragments are suitable unlike the conventional barcoding regions (Taberlet et al. 2007; Riaz et al. 2011). During eDNA barcoding analysis, the target genes readily form clusters named as molecular operational taxonomic units (MOTUs) if the taxons are unknown or the DNA database is low. eDNA metabarcoding analysis of shorter barcode exhibits a higher taxonomic resolution up to species level at a situation with fewer species-rich group or a lower biodiversity group (Bienert et al. 2012; Thomsen et al. 2012).

In this chapter, the authors have discussed about some scopes of eDNA approaches in current ecological studies which are very much essential for present-day conservation management regulations, especially in the aquatic environment.

8.1 Definition of eDNA

The eDNA or environmental DNA can be defined as "genetic material obtained directly from environmental samples (soil, sediment, water and biological living matter, etc.) without any obvious signs of biological source material" (Taberlet et al. 2012; Thomsen and Willerslev 2015). Thus, environmental DNA or eDNA could be well-defined as a tactic to attain genetic materials such as DNA from environment samples like soil, water, sediment, etc. Later stage, this genetic information is used by different aspects of biological or ecological sciences. The eDNA is generally found in a complex structure, i.e.,mixture of different forms such as dissolved, particulate absorbed, intracellular, and intraorganellar (Mauvisseau et al. 2022).

The eDNA can be divided into two types microbial and macrobial based on their concept (Turner et al. 2015). The microbial eDNA comprises eDNA (intraorganismal and extraorganismal) from microorganisms such as bacteria. Environmental samples contain plenty of intraorganismal DNA and extraorganismal eDNA. But in case of large animals the chances of getting live microbes are less in comparison to microbial DNA (Thomsen and Willerslev 2015). eDNA of larger animals (Macrobial eDNA) enters the environment through the shed, excreted material, and dead materials, but the membrane-bound is released to the environment through the activities of microbes, whereas membrane less or extracellular eDNA released by the larger animals spontaneously. This extracellular eDNA is analogous to microbial eDNA and tends to accumulate more in sediment than in the water column in the aquatic environment.

8.2 eDNA Vs Traditional Methods

eDNA, one of the non-invasive survey methods, which hold good as per the animal ethics concern, unlike the traditional methods the target species (here fishes as we are considering the aquatic environment) are neither disturbed nor caught for identification purposes. This is one of the most advanced and reliable and more importantly it shows similar types of results to the traditional ones (Takahara et al. 2012). Besides, its highly sensitive detection also identifies the species if they are less in number. Occasionally, it also failed to detect target species that were known to persist (Thomsen et al. 2012). Hopefully, the continuous development of scientific technology especially sensibility and detection systems will enhance eDNA-related studies, especially in the fields of new detection. In the future, this technology will be helpful while studying the real-time presence of migratory species, great crested newt (species that require an aquatic and terrestrial area for their life history or life cycle), life stages of species, etc. (Thomsen et al. 2012). The eDNA analysis will be helpful to study the larvae or juveniles who inhabit the pond during their tenure inside the pond when the traditional approach is not sufficient.

Besides its non-invasive or non-destructive technique, eDNA studies are well known as a more economically feasible technique and the weather-related phenomenon does not create any hindrances. It substantially reduces the sampling effort and has more cost benefits than the traditional methods, especially during the distribution study of rare or threatened species (Jerde et al. 2011).

This eDNA-based technology does not follow the *one type fit for all situations*. The analysis of eDNA-based study is easier in stagnant water (lake, ponds, and reservoirs) compared to flowing water. Although many authors have performed the eDNA-based study in running water (Goldberg et al. 2011; Jerde et al. 2011; Minamoto et al. 2012). Sometimes the traditional methods give reliable results during the life history study or even if we are concerned to find out the live organisms in a particular environment. Thus, for species monitoring studies both traditional and contemporary techniques (eDNA) are required.

In a nutshell, the advantages of eDNA over the traditional ones are non-invasive method, highly sensitive, less expensive,and independent of weather phenomenon.

8.3 Persistence of eDNA

eDNA is very fragile and its persistence varies with different habitats ranging from a few hours in freshwater habitats to thousands of years in permafrost and fossil ice (Willerslev et al. 2004). Sometimes, the old eDNA samples (1 day to around 1 month after detachment of the animal body) are undetectable by PCR. Barnes et al. (2014) elaborated on the eDNA degradation phenomenon and emphasized the activities of microbes and their enzymes on eDNA and identified that environmental factors like biological oxygen demand (BOD), chlorophyll concentration, total eDNA concentration, and pH of water and sediment have a strong effect on its degradation.

The factors influencing the degradation or persistence of eDNA in the environment can be categorized into two: (1) Biotic and (2) Abiotic.

***Biotic*:** Various biotic factors like intraspecific variation in DNA production, stress, metabolism and size, distribution and density, feeding and diet, and reproduction and migration influence the eDNA concentration.

Intraspecific variation: Intra-specific variations are observed in DNA production and shedding rates due to individual or species-specific issues although they have lived in the same habitat (Kelly et al. 2014). Species-specific factors like the sex of the individual, basal metabolism, and stress could have influencing factors for this (Nevers et al. 2018; Horiuchi et al. 2019).

Stress: Under the stressed condition fish sheds more eDNA immediately in response to the physiological stress, later on the rate is reduced afterward during the acclimatization process (Sassoubre et al. 2016; Nevers et al. 2018).

Metabolic rate and size: Higher metabolic rate in response to high temperature (Gillooly et al. 2001) leads to more rupture of upper skin, higher metabolites production, and other eruptions or excretions eventually more eDNA concentration in the environment (Maruyama et al. 2014; Chabot et al. 2016; Lacoursière-Roussel et al. 2016).

Distribution and density: The eDNA concentration is also affected by the patchy distribution of fishes; sampling across the tempo-spatial scale enables a good concentration of eDNA (Moyer et al. 2014; Furlan et al. 2016). A heterogeneously distributed population tends to have less chance of eDNA concentration in comparison to a moderately dense population (Ghosal et al. 2018). Fish density and eDNA concentration do not follow a consistent linear relationship always (Doi et al. 2017; Ghosal et al. 2018; Coulter et al. 2019).

Feeding and diet: Some reports have advocated that species performing an active feeding habit likely produce more eDNA than their lesser counterparts (Ghosal et al. 2018). The diet composition as well as the frequency of feeding affect the shedding and subsequently eDNA concentrations in the environment (Klymus et al. 2015).

Reproduction and migration: In aquatic systems, the presence of fish eDNA concentrations is related to their different life stages like breeding, movement, or migration (Rourke et al. 2022). The spawning and migration (migration related to breeding purpose) activities are associated with a large congregation of juveniles and adult fishes resulting in hyper activities and release of gametes; eventually, it leads to more eDNA concentration (Spear et al. 2015; Plough 2018; Bracken et al. 2019). Bylemans et al. (2017) observed a higher ratio of nuclear-to-mitochondrial eDNA of Macquarie perch (*Macquaria australasica*) during the spawning period and Bracken et al. (2019) also observed more sea lamprey (*Petromyzon marinus*) eDNA concentration in river water during the spawning period.

Abiotic and other factors: The environmental (nonbiological) factors like water flow, water temperature, water depth, environmental DNA decay, and other factors like methods of collecting, extracting, and amplifying eDNA and frequency and methods of eDNA sampling are responsible for eDNA concentrations in the aquatic environment. These factors exhibit an intrinsic complex interaction on general ecology, thus on the eDNA shedding rate and concentration.

Water flow: It is considered as the prime abiotic factor responsible for eDNA concentration. The water flow is the sole factor for the dispersion of eDNA from its source and spatial abundance along the flow path of system and it also affects the production, degradation, dilution, and deposition of eDNA (Thomsen et al. 2016; Hansen et al. 2018; Fukaya et al. 2021). Based on the flow regime, numerous studies have been undertaken to showcase the influence of water flow pattern on eDNA concentration of aquatic species, e.g., water flow effect on caged Atlantic salmon (*Salmo salar*) (Wood et al. 2020), and Japanese eel (*Anguilla japonica*) capture sites (Itakura et al. 2020). Various authors also observed the decline of eDNA concentration away from its source and no such accumulation pattern in the lower area or estuary in the lotic systems (Tillotson et al. 2018; Deutschmann et al. 2019; Wood et al. 2020).

Water temperature: Temperature has an intrinsic effect on species distribution, seasonal abundance, biology, and ecology (Takahara et al. 2012) and so on its eDNA concentration and decay (Lacoursière-Roussel et al. 2016). Many authors have studied many aquaria-based experiments on the effect of temperature on fish eDNA viz. brook charr (*Salvelinus fontinalis*) (Lacoursière-Roussel et al. 2016), common carp eDNA (Takahara et al. 2012), lake trout (*Salvelinus namaycush*) (Lacoursière-Roussel et al. 2016), and Ayu (*Plecoglossus altivelis*) (Doi et al. 2017).

Water depth: The significant effects of depth of water were observed in bigger water bodies like lakes and seas, while the shallow water bodies hardly differ in species composition and so as the environment DNA (Eichmiller et al. 2014; Moyer et al. 2014; Hänfling et al. 2016; Klobucar et al. 2017; Murakami et al. 2019). Further, Yamamoto et al. (2016) opined that the surface and bottom water were not correlated and lower bound eDNA concentration was noticed in bottom water; thus, more emphasis should be given to the effect of water depth on eDNA.

Environmental DNA decay: Decay of eDNA is influenced by biotic agents like microbial presence and their activities and the enzymatic dynamics and abiotic factors like water salinity, temperature, pH, and sunlight (Schmelzle and Kinziger 2016; Andruszkiewicz et al. 2017; Hansen et al. 2018). Apart from ultraviolet (UV) rays are also responsible for decaying the eDNA sample (Lindahl 1993). The rate of decay is more in marine than freshwater habitats (Sassoubre et al. 2016). This decay of eDNA vividly affects the estimation of abundance and biomass (Sassoubre et al. 2016; Nevers et al. 2018; Coulter et al. 2019). The decaying of eDNA many times affects the abundance and biomass estimation (Schmelzle and Kinziger 2016; Perez et al. 2017).

Methods of sampling, extracting, and amplifying eDNA: Apart from the intrinsic environmental factors, the assessment of eDNA is also affected by analytical procedures like sample collection, DNA extraction procedures, preservations, primer designing, and PCR amplification (Knudsen et al. 2019; Kumar et al. 2020; Thalinger et al. 2021). For example, the filtration method is generally preferred in comparison to other methods like centrifugation and precipitation, as these methods are only suitable in controlled environment studies (as the water volume is less and eDNA concentration very high) (Maruyama et al. 2014; Klymus et al. 2015; Piggott 2016). Many authors have compared the effect of different capture and extraction methods on biomass studies such as brook charr (Lacoursière-Roussel et al. 2016), common carp (Eichmiller et al. 2016), carp Macquarie perch (Piggott 2016), and marine fishes (Guivas and Brammell 2020). Also, the sampling should be carried out on spatio-temporal scale for proper representation of eDNA for biomass and other studies (Salter et al. 2019; Shelton et al. 2019).

8.4 Scopes of eDNA in Conservation Biology

In conservation biology, eDNA offers a great scope and due to the advanced technologies, the conservation biologist prefers it. Some important and useful eDNA-based applications are followed.

8.4.1 Estimation of Biomass

In the fields of community ecology and population study, estimation of biomass is one of the fundamental units to knowing the species distribution as well as their intra- or inter-species interactions. So as for the conservation study, it plays a crucial role in frame-working the management strategy (Begon and Townsend 2020). Species distribution can be estimated by knowing the absence/presence information of species. Traditional survey methods like netting, electrofishing, and acoustic surveys are generally used for biomass estimation, which is sometimes found to be tricky and expensive, and invasive type. In this regard, e-DNA techniques show a better opportunity for biomass estimation. The concentration or density of eDNA is correlated with the abundance of organisms and their life stages (Goldberg et al. 2011; Thomsen et al. 2012; Pilliod et al. 2013). Many others have estimated the biomass by estimating the eDNA concentration in the environment such as biomass estimation of two amphibians spadefoot toad and great crested newt (Thomsen et al. 2012), common carp (Takahara et al. 2012), and carp biomass in a lagoon ecosystem (Pilliod et al. 2013). Pilliod et al. (2013) have also developed a general linear model for the estimation of biomass. The eDNA concentration in water is also influenced by water temperature, number, size, and behavior of fishes (Pilliod et al. 2014).

8.4.2 Species Composition

In the aquatic environment, the species composition study is more difficult than the terrestrial areas as the aquatic organisms are not easily assessable. So, the species composition study is much more exhaustive and expensive and the operation is limited to certain areas. Whereas, the eDNA-based NGS or next-generation sequencing helps the biologist to perform the species composition or the community study in an easier way overcoming all the above-mentioned difficulties. Thomsen et al.

(2012) have designed 6 primers for the identification of different aquatic organisms such as amphibians (Pelobates, Triturus) and fish (Tinca, Perca), etc. They had got 100% identical species, that have encountered through the traditional methods. Besides, they got more samples of species that were inhabiting adjacent to the aquaria. They have also demonstrated the NGS in marine environments and identified 15 species, but the results of the traditional method had only identified 9 species. This eDNA-based NGS study is fairly helpful for species composition and eventually, it will help the following thrust areas of ecological conservation and management.

1. Monitoring of species composition in different water bodies (lotic and lentic).
2. Monitoring the water pollution or restoration by knowing the indicator species through eDNA analysis.
3. Monitoring of restoration or recovery situations of some catastrophic events like oil spill.

8.4.3 Invasive Species

We all are aware about the impact of alien or invasive species on native fauna as well as the ecosystem. Early detection and their eradication at their low-density distribution is an effective tool for the invasive species management regime (Hulme 2006). At this juncture, the use of traditional techniques might be a shortfall for early detection. Thus, backing the potential of eDNA studies, it should be considered the best strategy for the management of alien species. Management of invasive Asian carps in waterways of the USA is a noted example of the use of eDNA technology for invasive species management (Jerde et al. 2011, 2013; Mahon et al. 2013b). Besides, other examples of eDNA for invasive detection include the American bullfrog in France (Ficetola et al. 2008; Dejean et al. 2011, 2012), the Bluegill sunfish in Japan (Takahara et al. 2015), the New Zealand mud snail in the Portneuf River, Idaho (Goldberg et al. 2011), *Cyprinus carpio* in Australia (Furlan et al. 2019), Crayfish in Norwegian freshwater habitat (Strand et al. 2019), and the Burmese python in southern Florida (Piaggio et al. 2014). The repeated detection indicates the presence of live species in the system followed by innovative screening methods such as microfluidic carbon nanotube platform and laser transmission spectroscopy (Mahon et al. 2011, 2013b) are used for surveillance cum eradications. We all aware that ballast water is a source of introduction of invasive species (Chen et al. 2022). For the detection of alien species in ballast water, Mohan et al. have developed an innovative microfluidic carbon nanotube platform, that is used to check the presence of species like the zebra mussel, the quagga mussel and the Chinese mitten crab in ballast water (Mahon et al. 2011). These alien species are a potential threat for the Laurentian Great Lakes, to control their effective non-introduction, the ballast water was gone through this screening procedure before port entry. In this screening process, species-specific primers are used for amplification of single-standard DNA within a chip loaded with 26-bp-long DNA oligonucleotides with reverse complement for the amplification purpose. After the application of the alternating current (AC), a rising impedance value has noticed when the DNA continues to bind with oligonucleotides; this technique is helping for the real-time detection of exotic species in ballast water. The later eDNA-based screening technology includes a hybridization of PCR amplified product with the nanoparticles of species-specific oligonucleotide probes and used for the detection of the invasive freshwater quagga mussel (Mahon et al. 2013a).

8.4.4 Rare and Threatened Species

Globally, about 5900 freshwater and marine animal species are categorized as endangered by IUCN (IUCN redlist 2022). Backing to the superiority of eDNA technology from the traditional methods, it is suitable to use this technology in data collection for rare and threatened species due to their low abundance, small number, and conservation-related restrictions (Thomsen et al. 2012). Also, this technique was successfully demonstrated in a few cases such as for the eastern hellbender, an amphibian of conservation concern (Olson et al. 2012), the blind cave eel *Ophisternon candidum* endemic to Australia (White et al. 2020), Sawfish in Australia (Cooper et al. 2021), Chinook salmon (*Oncorhynchus tshawytscha*) in the USA (Shelton et al. 2019), etc.

8.4.5 Ancient Environments and Extinctions

Ancient DNA (a DNA) study has started in the year 1984 from the quagga (an extinct mussel), with the development of science the study on a DNA has increased and now scientists are able to do whole genome sequence of willy mammoths, cave bear, human races like Vikings to Neanderthals covering a temporal frame of half million years ago (Orlando et al. 2021). Analysis of a DNA is a mammoth task as the DNA samples are degraded significantly with very small DNA fragments (Willerslev and Cooper 2005). With the help of NGS technology we are able to acquire such huge data across the temporal scale, which are helpful for studying the genetic ancestries of animals, fungi and plants, paleomicrobiology as well as to know the population study and the evolution etc. This ancient data (*a DNA* or *h DNA*) gives baseline information on biodiversity, through which we are able to distinguish between "now" and "then" of biodiversity and this information is essential for guiding the conservation measures and policies (Jensen et al. 2022). Not only the phylogeny and genetic variation, a *DNA* also notify about the dietary pattern, social life, infectious diseases, virulence, adaption to gradual or episodic changes, etc. (Orlando et al. 2021). In paleological studies, aquatic sediment (the *sedaDNA*) is one of the best suitable options to understand the past life (ancient biota including both the macro and microorganisms and their complex communities) (Capo et al. 2022). Besides, knowledge of such past ecological changes through these ancient studies is helping us to understand the changes rendered by climate change or warming effects (Fig. 8.2).

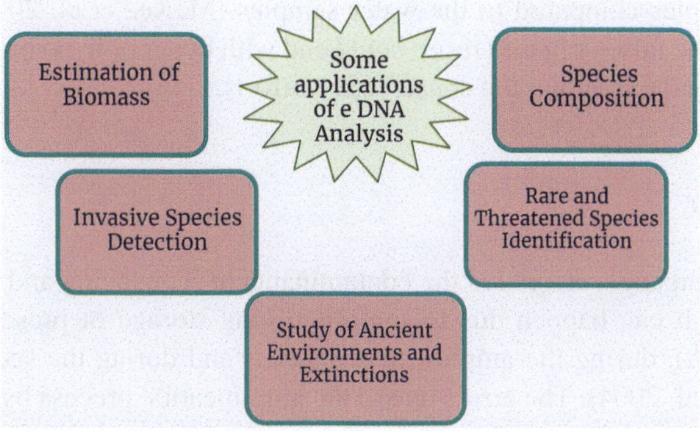

Fig. 8.2 Some important applications of eDNA analysis in the aquatic environment

8.5 Pitfalls and Challenges for eDNA

Despite its practical utility and remarkable outlook for conservation as well as ecology, eDNA analysis has its own shortcomings and the researchers must be cautious about these.

8.5.1 Contamination

The most common and menacing factor of eDNA-related work is contamination. The sample contamination can occur at any point of sample analysis, whether it is at the beginning, the sample collection process or at the later stages during the laboratory analysis. Contaminated samples give false result, which is a matter of concern. During sampling, repeated or continuous sampling or if sampling not follow the proper sampling procedure could be the reasons for cross contaminations. At the laboratory, repeated use of PCR and not following the correct laboratory protocols could end up with freely roaming billions of DNA products in the laboratory and causing cross-contamination. For the NGS technologies even it is more dangerous. Though cross-contamination is an unavoidable thing in the laboratory, but following strict lab protocols (even the sampling procedure during sampling) and decontamination procedures before and after PCR work will reduce cross-contamination drastically (Willerslev and Cooper 2005; Champlot et al. 2010).

8.5.2 Inhibition

The enzyme-inhibiting substances like humic acid and other humic components are stubborn particles in eDNA samples and they are usually co-extracted with the samples. They inhibit the *Taq* polymerase (used in PCR reactions to amplify DNA) activity resulting in a biased result. The inhibition is typically observed in soil and sediment samples compared to the water samples (McKee et al. 2015; Sigsgaard et al. 2015). The false or biased result could end with hyper or hypo biomass estimation, subsequently affecting the conservation efforts.

8.5.3 Error

In layman's language, errors are the contaminants in sequencing and resulting in a biased result. It can happen due to sampling, long storage of preserved samples (DNA or RNA), during the amplification process and during the sequencing also (Willerslev et al. 2004). The errors due to the amplification process by PCR include point mutations and formed chimeric molecules (Acinas et al. 2005). These errors masked the biodiversity signals during animal biodiversity assessment (Sefc et al. 2007). During microbial biodiversity studies, the sequencing errors overestimate the biodiversity (Kunin et al. 2010). So, the raw sequence must go through screening to check the false positive errors for getting the taxon list.

8.5.4 Database

Reference database plays a crucial role for identification of the DNA sequences of the environmental sample through the process named as *blasting*. Some widely used databases are ncbi database(www.ncbi.nlm.nih.gov/genbank), (www.boldsystems.org) . Some DNA barcode databases are for fish (www.fishbol.org), for mammal (www.mammaliabol.org) and for bird species, (www.barcodingbirds.org) *etc*. However, the databases are still in suboptimal condition; so backing to their popularity and importance, the DNA reference databases should focus on complete mitochondrial or even nuclear genome or whole genome for tits broader applicability.

8.6 Single Species Detection Vs. DNA Metabarcoding

For the identification of species, two types of species detection methods, single species detection and multi-species detection technology, are used. Where the former is used to identify single or specific species through the species-specific designed primer followed by PCR or q PCR and the latter uses the general primer and the multi-species detection carried out by the NGS (Takahara et al. 2012). Although, the former possesses a higher specificity and sensitivity and higher quantification ability, its identification detectability is limited to single species only. So, for a highly diverse ecosystem, its applicability is limited due to the deficiency of DNA extract for multiple reaction and the expense of more money. Besides, qPCR-related eDNA studies has some limitations such as cycle threshold (Ct) cut-off values, percentage of positive replicates for defining true positive values from background, proper treatment of negative qPCR replicates, the requirement of ample qPCR replicates throughout the study, etc. (Ellison et al. 2006; Bustin et al. 2009).

Although the DNA metabarcoding methods are cost-effective and powerful tools for mass DNA sequence, its generic primer shows specific affinity towards particular sequences, thus not amplified properly (Deagle et al. 2014). As we know NGS are complex sequencing process and required to be filtered for removal of the errors or low abundant sequences. This removal process is known as denoising and it is an important part of NGS to provide accurate estimation of the biodiversity.

8.6.1 Interpretation of Results

Interpretation of the final result is the most crucial one. eDNA detection has some limitations to distinguish between the living and the dead organisms, even the life stages and hybrid species. The mtDNA which carries maternal genome could establish the maternal or ancestor's lineage and also help the hybrid species interpretation. Nowadays ecologists use the occupancy models for monitoring the ecosystems (Schadt et al. 2010).

8.7 A Way Forward of eDNA

eDNA-based analysis is rapidly fetching importance in the field of ecological studies. For the conservation tactics, where the real-time monitoring or periodical monitoring is an integral part is strengthened by the eDNA technologies. This cost-effective, rapid, reliable and sensitive, NGS-based eDNA technique offers a promising management regime. Still, its applicability needs to be enhanced through the advancement and modification of existing practices.

1. Development of a quick on-site detection kit to be used in sampling sites for rapid and real-time detection of organisms.
2. Use of alternative methods like loop-mediated isothermal amplification or recombinase polymerase amplification for faster processing and lesser requirement of high-end equipment.
3. Improvement of the methodologies for minimal environment inhibition during analysis.
4. Methods development for minimizing the errors and to know the origin of eDNA whether from epithelial cells or intestinal cell or from excreted materials.
5. Substantial development of Methodology and technology to distinguish between the living and dead and different life stages also.
6. eDNA-based tempo and spatial distribution study in different habitats to indulge the species abundance study for effective biodiversity monitoring and conservation.

8.8 Conclusion

eDNA is a powerful addition to ecological studies, even the integration of the NGS to the eDNA makes it more powerful. Through the NGS technology, ecology managers are excavating new avenues in ecological studies whether it is monitoring programs (species richness, population abundance, invasive species monitoring, and rare threatened species status) or multi-species invention. This eDNA approach should complement the traditional monitoring methodology rather than replace it. A person with sound ecological knowledge (including grassroots-level knowledge) is able to provide a meaningful interpretation of results followed by recommendations. We should use this technology for the welfare of planet earth and its biodiversity.

References

Acinas SG, Sarma-Rupavtarm R, Klepac-Ceraj V, Polz MF (2005) PCR-induced sequence artifacts and bias: insights from comparison of two 16S rRNA clone libraries constructed from the same sample. Appl Environ Microbiol 71:8966

Andruszkiewicz EA, Sassoubre LM, Boehm AB (2017) Persistence of marine fish environmental DNA and the influence of sunlight. PLoS one 12:e0185043

Baird DJ, Hajibabaei M (2012) Biomonitoring 2.0: a new paradigm in ecosystem assessment made possible by next-generation DNA sequencing. Mol Ecol 21:2039

Baldwin CC, Collette BB, Parenti LR, Smith DG, Springer VG (1996) Methods and techniques of underwater research. In: Proceedings of the American Academy of underwater sciences scientific diving symposium. Smithsonian Institution, Washington

Barnes MA, Turner CR, Jerde CL, Renshaw MA, Chadderton WL, Lodge DM (2014) Environmental conditions influence eDNA persistence in aquatic systems. Environ Sci Technol 48:1819–1827

Begon M, Townsend CR (2020) Ecology: from individuals to ecosystems. Wiley, New York

Bellemain E, Carlsen T, Brochmann C, Coissac E, Taberlet P, Kauserud H (2010) ITS as an environmental DNA barcode for fungi: an in silico approach reveals potential PCR biases. BMC Microbiol 10:1–9

Bienert F, De Danieli S, Miquel C, Coissac E, Poillot C, BRUN J, Taberlet P (2012) Tracking earthworm communities from soil DNA. Mol Ecol 21:2017–2030

Bracken FSA, Rooney SM, Kelly-Quinn M, King JJ, Carlsson J (2019) Identifying spawning sites and other critical habitat in lotic systems using eDNA "snapshots": a case study using the sea lamprey Petromyzon marinus L. Ecol Evol 9:553–567

Bunce M, Szulkin M, Lerner HRL, Barnes I, Shapiro B, Cooper A, Holdaway RN (2005) Ancient DNA provides new insights into the evolutionary history of New Zealand's extinct giant eagle. PLoS Biology 3:e9

Bustin SA, Benes V, Garson JA, Hellemans J, Huggett J, Kubista M, Mueller R, Nolan T, Pfaffl MW, Shipley GL (2009) The MIQE guidelines: M inimum I nformation for publication of Q uantitative real-time PCR Experiments. Clin Chem 55:611

Bylemans J, Furlan EM, Hardy CM, McGuffie P, Lintermans M, Gleeson DM (2017) An environmental DNA-based method for monitoring spawning activity: a case study, using the endangered Macquarie perch (Macquaria australasica). Methods Ecol Evol 8:646–655

Capo E, Monchamp M, Coolen MJL, Domaizon I, Armbrecht L, Bertilsson S (2022) Environmental paleomicrobiology: using DNA preserved in aquatic sediments to its full potential. Environ Microbiol 24:2201

Chabot D, Steffensen JF, Farrell AP (2016) The determination of standard metabolic rate in fishes. J Fish Biol 88:81–121

Champlot S, Berthelot C, Pruvost M, Bennett EA, Grange T, Geigl E-M (2010) An efficient multistrategy DNA decontamination procedure of PCR reagents for hypersensitive PCR applications. PloS one 5:e13042

Chen Y-S, Kang C-K, Liu T-K (2022) Ballast water management strategy to reduce the impact of introductions by utilizing an empirical risk model. Water 14:981

Cooper MK, Huerlimann R, Edmunds RC, Budd AM, Le Port A, Kyne PM, Jerry DR, Simpfendorfer CA (2021) Improved detection sensitivity using an optimal eDNA preservation and extraction workflow and its application to threatened sawfishes. Aquat conserv 31:2131–2148

Coulter DP, Wang P, Coulter AA, Van Susteren GE, Eichmiller JJ, Garvey JE, Sorensen PW (2019) Nonlinear relationship between silver carp density and their eDNA concentration in a large river. PloS One 14:e0218823

Daan N (2001) The IBTS database: a plea for quality control

Deagle BE, Jarman SN, Coissac E, Pompanon F, Taberlet P (2014) DNA metabarcoding and the cytochrome c oxidase subunit I marker: not a perfect match. Biol Lett 10:20140562

Dejean T, Valentini A, Duparc A, Pellier-Cuit S, Pompanon F, Taberlet P, Miaud C (2011) Persistence of environmental DNA in freshwater ecosystems. PloS One 6:e23398

Dejean T, Valentini A, Miquel C, Taberlet P, Bellemain E, Miaud C (2012) Improved detection of an alien invasive species through environmental DNA barcoding: the example of the American bullfrog Lithobates catesbeianus. J Appl Ecol 49:953–959

Deutschmann B, Müller A-K, Hollert H, Brinkmann M (2019) Assessing the fate of brown trout (Salmo trutta) environmental DNA in a natural stream using a sensitive and specific dual-labelled probe. Sci Total Environ 655:321–327

Díaz S, Fargione J, Chapin FS III, Tilman D (2006) Biodiversity loss threatens human well-being. PLoS Biology 4:e277

Dirzo R, Ceballos G, Ehrlich PR (2022) Circling the drain: the extinction crisis and the future of humanity. Philos Trans R Soc B 377:20210378

Doi H, Inui R, Akamatsu Y, Kanno K, Yamanaka H, Takahara T, Minamoto T (2017) Environmental DNA analysis for estimating the abundance and biomass of stream fish. Freshwater Biology 62:30–39

Eichmiller JJ, Bajer PG, Sorensen PW (2014) The relationship between the distribution of common carp and their environmental DNA in a small lake. PloS One 9:e112611

Eichmiller JJ, Miller LM, Sorensen PW (2016) Optimizing techniques to capture and extract environmental DNA for detection and quantification of fish. Mol Ecol Resour 16:56–68

Ellison SLR, English CA, Burns MJ, Keer JT (2006) Routes to improving the reliability of low level DNA analysis using real-time PCR. BMC Biotechnol 6:1–11

Ficetola GF, Miaud C, Pompanon F, Taberlet P (2008) Species detection using environmental DNA from water samples. Biol Lett 4:423–425

Fukaya K, Murakami H, Yoon S, Minami K, Osada Y, Yamamoto S, Masuda R, Kasai A, Miyashita K, Minamoto T (2021) Estimating fish population abundance by integrating quantitative data on environmental DNA and hydrodynamic modelling. Molecular Ecol 30:3057–3067

Furlan EM, Gleeson D, Hardy CM, Duncan RP (2016) A framework for estimating the sensitivity of eDNA surveys. Mol Ecol Resour 16:641–654

Furlan EM, Gleeson D, Wisniewski C, Yick J, Duncan RP (2019) eDNA surveys to detect species at very low densities: a case study of European carp eradication in Tasmania, Australia. J Appl Ecol 56:2505–2517

Ghosal R, Eichmiller JJ, Witthuhn BA, Sorensen PW (2018) Attracting common carp to a bait site with food reveals strong positive relationships between fish density, feeding activity, environmental DNA, and sex pheromone release that could be used in invasive fish management. Ecol Evol 8:6714–6727

Gillooly JF, Brown JH, West GB, Savage VM, Charnov EL (2001) Effects of size and temperature on metabolic rate. Science 293:2248–2251

Goldberg CS, Pilliod DS, Arkle RS, Waits LP (2011) Molecular detection of vertebrates in stream water: a demonstration using Rocky Mountain tailed frogs and Idaho giant salamanders. PloS One 6:e22746

Guivas RA, Brammell BF (2020) Use of environmental DNA to determine fantail darter (Etheostoma flabellare) density in a laboratory setting: effects of biomass and filtration method. Int J Zool 2020:4731686

Hänfling B, Lawson Handley L, Read DS, Hahn C, Li J, Nichols P, Blackman RC, Oliver A, Winfield IJ (2016) Environmental DNA metabarcoding of lake fish communities reflects long-term data from established survey methods. Mol Ecol 25:3101–3119

Hansen BK, Bekkevold D, Clausen LW, Nielsen EE (2018) The sceptical optimist: challenges and perspectives for the application of environmental DNA in marine fisheries. Fish Fisheries 19:751–768

Hebert PDN, Cywinska A, Ball SL, DeWaard JR (2003) Biological identifications through DNA barcodes. Proc Biol Sci 270:313–321

Hopkins GW, Freckleton RP (2002) Declines in the numbers of amateur and professional taxonomists: implications for conservation. In: Animal conservation forum, vol 5. Cambridge University Press, Cambridge, pp 245–249

Horiuchi T, Masuda R, Murakami H, Yamamoto S, Minamoto T (2019) Biomass-dependent emission of environmental DNA in jack mackerel Trachurus japonicus juveniles. J Fish Biol 95:979–981

Hulme PE (2006) Beyond control: wider implications for the management of biological invasions. J Appl Ecol 43:835–847

Itakura H, Wakiya R, Sakata MK, Hsu H-Y, Chen S-C, Yang C-C, Huang Y-C, Han Y-S, Yamamoto S, Minamoto T (2020) Estimations of riverine distribution, abundance, and biomass of Anguillid eels in Japan and Taiwan using environmental DNA analysis. Zool Stud 59:e17

IUCN (2022) The IUCN red list of threatened species. Version 2022-2. https://www.iucnredlist.org. Acessed 22 Dec 2022

Jensen EL, Díez-del-Molino D, Gilbert MTP, Bertola LD, Borges F, Cubric-Curik V, de Navascués M, Frandsen P, Heuertz M, Hvilsom C (2022) Ancient and historical DNA in conservation policy. Trends Ecol Evol 37:420

Jerde CL, Mahon AR, Chadderton WL, Lodge DM (2011) "Sight-unseen" detection of rare aquatic species using environmental DNA. Conserv Lett 4:150–157

Jerde CL, Chadderton WL, Mahon AR, Renshaw MA, Corush J, Budny ML, Mysorekar S, Lodge DM (2013) Detection of Asian carp DNA as part of a Great Lakes basin-wide surveillance program. Can J Fish Aquat Sci 70:522–526

Kelly RP, Port JA, Yamahara KM, Martone RG, Lowell N, Thomsen PF, Mach ME, Bennett M, Prahler E, Caldwell MR (2014) Harnessing DNA to improve environmental management. Science 344:1455–1456

Klobucar SL, Rodgers TW, Budy P (2017) At the forefront: evidence of the applicability of using environmental DNA to quantify the abundance of fish populations in natural lentic waters with additional sampling considerations. Can J Fish Aquat Sci 74:2030–2034

Klymus KE, Richter CA, Chapman DC, Paukert C (2015) Quantification of eDNA shedding rates from invasive bighead carp Hypophthalmichthys nobilis and silver carp Hypophthalmichthys molitrix. Biol Conserv 183:77–84

Knudsen SW, Ebert RB, Hesselsøe M, Kuntke F, Hassingboe J, Mortensen PB, Thomsen PF, Sigsgaard EE, Hansen BK, Nielsen EE (2019) Species-specific detection and quantification of environmental DNA from marine fishes in the Baltic Sea. J Exp Mar Biol Ecol 510:31–45

Kumar G, Eble JE, Gaither MR (2020) A practical guide to sample preservation and pre-PCR processing of aquatic environmental DNA. Mol Ecol Resou 20:29–39

Kunin V, Engelbrektson A, Ochman H, Hugenholtz P (2010) Wrinkles in the rare biosphere: pyrosequencing errors can lead to artificial inflation of diversity estimates. Environ Microbiol 12:118–123

Lacoursière-Roussel A, Côté G, Leclerc V, Bernatchez L (2016) Quantifying relative fish abundance with eDNA: a promising tool for fisheries management. J Appl Ecol 53:1148–1157

Lindahl T (1993) Instability and decay of the primary structure of DNA. Nature 362:709–715

Lydolph MC, Jacobsen J, Arctander P, Gilbert MTP, Gilichinsky DA, Hansen AJ, Willerslev E, Lange L (2005) Beringian paleoecology inferred from permafrost-preserved fungal DNA. Appl Environ Microbiol 71:1012–1017

Mahon AR, Barnes MA, Senapati S, Feder JL, Darling JA, Chang H-C, Lodge DM (2011) Molecular detection of invasive species in heterogeneous mixtures using a microfluidic carbon nanotube platform. PLoS One 6:e17280

Mahon AR, Barnes MA, Li F, Egan SP, Tanner CE, Ruggiero ST, Feder JL, Lodge DM (2013a) DNA-based species detection capabilities using laser transmission spectroscopy. J R Soc Interface 10:20120637

Mahon AR, Jerde CL, Galaska M, Bergner JL, Chadderton WL, Lodge DM, Hunter ME, Nico LG (2013b) Validation of eDNA surveillance sensitivity for detection of Asian carps in controlled and field experiments. PloS One 8:e58316

Maruyama A, Nakamura K, Yamanaka H, Kondoh M, Minamoto T (2014) The release rate of environmental DNA from juvenile and adult fish. PLoS One 9:e114639

Mauvisseau Q, Harper LR, Sander M, Hanner RH, Kleyer H, Deiner K (2022) The multiple states of environmental DNA and what is known about their persistence in aquatic environments. Environ Sci Technol 56:5322–5333

McKee AM, Spear SF, Pierson TW (2015) The effect of dilution and the use of a post-extraction nucleic acid purification column on the accuracy, precision, and inhibition of environmental DNA samples. Biol Conserv 183:70–76

Minamoto T, Yamanaka H, Takahara T, Honjo MN, Kawabata Z (2012) Surveillance of fish species composition using environmental DNA. Limnology 13:193–197

Moyer GR, Diaz-Ferguson E, Hill JE, Shea C (2014) Assessing environmental DNA detection in controlled lentic systems. PloS One 9:e103767

Murakami H, Yoon S, Kasai A, Minamoto T, Yamamoto S, Sakata MK, Horiuchi T, Sawada H, Kondoh M, Yamashita Y (2019) Dispersion and degradation of environmental DNA from caged fish in a marine environment. Fish Sci 85:327–337

Nevers MB, Byappanahalli MN, Morris CC, Shively D, Przybyla-Kelly K, Spoljaric AM, Dickey J, Roseman EF (2018) Environmental DNA (eDNA): a tool for quantifying the abundant but elusive round goby (Neogobius melanostomus). PloS One 13:e0191720

Nielsen KM, Johnsen PJ, Bensasson D, Daffonchio D (2007) Release and persistence of extracellular DNA in the environment. Environ Biosaf Res 6:37–53

Olson ZH, Briggler JT, Williams RN (2012) An eDNA approach to detect eastern hellbenders (Cryptobranchus a. alleganiensis) using samples of water. Wildlife Res 39:629–636

Orlando L, Allaby R, Skoglund P, Der Sarkissian C, Stockhammer PW, Ávila-Arcos MC, Fu Q, Krause J, Willerslev E, Stone AC (2021) Ancient DNA analysis. Nat Rev Method Prim 1:1–26

Perez CR, Bonar SA, Amberg JJ, Ladell B, Rees C, Stewart WT, Gill CJ, Cantrell C, Robinson AT (2017) Comparison of American fisheries society (AFS) standard fish sampling techniques and environmental DNA for characterizing fish communities in a large reservoir. North Am J Fish Manag 37:1010–1027

Piaggio AJ, Engeman RM, Hopken MW, Humphrey JS, Keacher KL, Bruce WE, Avery ML (2014) Detecting an elusive invasive species: a diagnostic PCR to detect B urmese python in F lorida waters and an assessment of persistence of environmental DNA. Mol Ecol Resour 14:374–380

Piggott MP (2016) Evaluating the effects of laboratory protocols on eDNA detection probability for an endangered freshwater fish. Ecol Evol 6:2739–2750

Pilliod DS, Goldberg CS, Arkle RS, Waits LP (2013) Estimating occupancy and abundance of stream amphibians using environmental DNA from filtered water samples. Can J Fish Aquat Sci 70:1123–1130

Pilliod DS, Goldberg CS, Arkle RS, Waits LP (2014) Factors influencing detection of eDNA from a stream-dwelling amphibian. Mol Ecol Resou 14:109–116

Plough LV, Ogburn MB, Fitzgerald CL, Geranio R, Marafino GA, Richie KD (2018) Environmental DNA analysis of river herring in Chesapeake Bay: a powerful tool for monitoring threatened keystone species. PLoS One 13:e0205578

Poinar HN, Hofreiter M, Spaulding WG, Martin PS, Stankiewicz BA, Bland H, Evershed RP, Possnert G, Paabo S (1998) Molecular coproscopy: dung and diet of the extinct ground sloth Nothrotheriops shastensis. Science 281:402–406

Riaz T, Shehzad W, Viari A, Pompanon F, Taberlet P, Coissac E (2011) ecoPrimers: inference of new DNA barcode markers from whole genome sequence analysis. Nucl Acids Res 39:e145–e145

Rourke ML, Fowler AM, Hughes JM, Broadhurst MK, DiBattista JD, Fielder S, Wilkes Walburn J, Furlan EM (2022) Environmental DNA (eDNA) as a tool for assessing fish biomass: a review of approaches and future considerations for resource surveys. Environ DNA 4:9–33

Salter I, Joensen M, Kristiansen R, Steingrund P, Vestergaard P (2019) Environmental DNA concentrations are correlated with regional biomass of Atlantic cod in oceanic waters. Commun Biol 2:1–9

Sassoubre LM, Yamahara KM, Gardner LD, Block BA, Boehm AB (2016) Quantification of environmental DNA (eDNA) shedding and decay rates for three marine fish. Environ Sci Technol 50:10456–10464

Schadt EE, Turner S, Kasarskis A (2010) A window into third-generation sequencing. Human Mol Genet 19:R227–R240

Schmelzle MC, Kinziger AP (2016) Using occupancy modelling to compare environmental DNA to traditional field methods for regional-scale monitoring of an endangered aquatic species. Mol Ecol Resour 16:895–908

Schoch CL, Seifert KA, Huhndorf S, Robert V, Spouge JL, Levesque CA, Chen W, Consortium FB, List FBCA, Bolchacova E (2012) Nuclear ribosomal internal transcribed spacer (ITS) region as a universal DNA barcode marker for fungi. Proc Natl Acad Sci 109:6241–6246

Sefc KM, Payne RB, Sorenson MD (2007) Single base errors in PCR products from avian museum specimens and their effect on estimates of historical genetic diversity. Conserv Genet 8:879–884

Shelton AO, Kelly RP, O'Donnell JL, Park L, Schwenke P, Greene C, Henderson RA, Beamer EM (2019) Environmental DNA provides quantitative estimates of a threatened salmon species. Biol Conserv 237:383–391

Shokralla S, Spall JL, Gibson JF, Hajibabaei M (2012) Next-generation sequencing technologies for environmental DNA research. Mol Ecol 21:1794–1805

Sigsgaard EE, Carl H, Møller PR, Thomsen PF (2015) Monitoring the near-extinct European weather loach in Denmark based on environmental DNA from water samples. Biol Conserv 183. Elsevier:46–52

Spear SF, Groves JD, Williams LA, Waits LP (2015) Using environmental DNA methods to improve detectability in a hellbender (Cryptobranchus alleganiensis) monitoring program. Biol Conserv 183:38–45

Strand DA, Johnsen SI, Rusch JC, Agersnap S, Larsen WB, Knudsen SW, Møller PR, Vrålstad T (2019) Monitoring a Norwegian freshwater crayfish tragedy: eDNA snapshots of invasion, infection and extinction. J Appl Ecol 56:1661–1673

Taberlet P, Coissac E, Pompanon F, Gielly L, Miquel C, Valentini A, Vermat T, Corthier G, Brochmann C, Willerslev E (2007) Power and limitations of the chloroplast trn L (UAA) intron for plant DNA barcoding. Nucl Acid Res 35:e14

Taberlet P, Coissac E, Hajibabaei M, Rieseberg LH (2012) Environmental DNA. Mol Ecol 21:1789

Takahara T, Minamoto T, Yamanaka H, Doi H, Kawabata Z (2012) Estimation of fish biomass using environmental DNA. PloS One 7:e35868

Takahara T, Minamoto T, Doi H (2015) Effects of sample processing on the detection rate of environmental DNA from the common carp (Cyprinus carpio). Biol Conserv 183:64–69

Thalinger B, Deiner K, Harper LR, Rees HC, Blackman RC, Sint D, Traugott M, Goldberg CS, Bruce K (2021) A validation scale to determine the readiness of environmental DNA assays for routine species monitoring. Environ DNA 3:823–836

Thomsen PF, Willerslev E (2015) Environmental DNA–an emerging tool in conservation for monitoring past and present biodiversity. Biol Conserv 183:4–18

Thomsen PF, Kielgast JOS, Iversen LL, Wiuf C, Rasmussen M, Gilbert MTP, Orlando L, Willerslev E (2012) Monitoring endangered freshwater biodiversity using environmental DNA. Mol Ecol 21:2565–2573

Thomsen PF, Møller PR, Sigsgaard EE, Knudsen SW, Jørgensen OA, Willerslev E (2016) Environmental DNA from seawater samples correlate with trawl catches of subarctic, Deepwater fishes. PloS One 11:e0165252

Tillotson MD, Kelly RP, Duda JJ, Hoy M, Kralj J, Quinn TP (2018) Concentrations of environmental DNA (eDNA) reflect spawning salmon abundance at fine spatial and temporal scales. Biol Conserv 220:1–11

Turner CR, Barnes MA, Xu CCY, Jones SE, Jerde CL, Lodge DM (2014) Particle size distribution and optimal capture of aqueous macrobial eDNA. Method Ecol Evol 5:676–684

Turner CR, Uy KL, Everhart RC (2015) Fish environmental DNA is more concentrated in aquatic sediments than surface water. Biol Conserv 183:93–102

Valiere N, Taberlet P (2000) Urine collected in the field as a source of DNA for species and individual identification. Mol Ecol 9:2150–2152

Wheeler QD, Raven PH, Wilson EO (2004) Taxonomy: impediment or expedient? Science 304:285

White NE, Guzik MT, Austin AD, Moore GI, Humphreys WF, Alexander J, Bunce M (2020) Detection of the rare Australian endemic blind cave eel (Ophisternon candidum) with environmental DNA: implications for threatened species management in subterranean environments. Hydrobiologia 847:3201–3211

Willerslev E, Cooper A (2005) Ancient DNA. Proc Royal Soc B Biol Sci 272:3–16

Willerslev E, Hansen AJ, Binladen J, Brand TB, Gilbert MTP, Shapiro B, Bunce M, Wiuf C, Gilichinsky DA, Cooper A (2003) Diverse plant and animal genetic records from Holocene and Pleistocene sediments. Science 300:791–795

Willerslev E, Hansen AJ, Rønn R, Brand TB, Barnes I, Wiuf C, Gilichinsky D, Mitchell D, Cooper A (2004) Long-term persistence of bacterial DNA. Curr Biol 14:R9–R10

Wood ZT, Erdman BF, York G, Trial JG, Kinnison MT (2020) Experimental assessment of optimal lotic eDNA sampling and assay multiplexing for a critically endangered fish. Environ DNA 2:407–417

Yamamoto S, Minami K, Fukaya K, Takahashi K, Sawada H, Murakami H, Tsuji S, Hashizume H, Kubonaga S, Horiuchi T (2016) Environmental DNA as a 'snapshot'of fish distribution: a case study of Japanese jack mackerel in Maizuru Bay, sea of Japan. PloS One 11:e0149786

Genomics Data Analysis Techniques in Bioinformatics

9

Smruti Priyambada Pradhan, Ajaya Kumar Rout, Epari Venkata Rao, and Sukanta Kumar Pradhan

Abstract

Genomic data analysis is very useful for understanding critical computational raw data. Using current genomic technology, a single biological experiment generates millions of data points. The generated data is straining bioinformatics pipelines due to the widespread use of these technologies. Because of this, our understanding of how the genome works is becoming less based on data and more on analysis. Only until *de-novo* whole-genome shotgun sequencing and annotation are achieved are a deep organizational study and comprehension of the whole landscape of a genome feasible. Genomic resources including high-density SNP genotyping arrays, whole genome reference sequences, and sequenced genotype are being created for a variety of projects. The scientific community will have access to these materials, which will provide them strong tools to use in identifying the genetic elements that control complex phenotypes.

Keywords

Genomics · Next-Generation Sequencing · Single nucleotide polymorphisms · Omics · De-novo

S. P. Pradhan · E. V. Rao
Department of Community Medicine, IMS & SUM Hospital, Bhubaneswar, India

A. K. Rout
Aquatic Environmental Biotechnology and Nanotechnology Division, ICAR-Central Inland Fisheries Research Institute, Barrackpore, Kolkata, West Bengal, India

S. K. Pradhan (✉)
Department of Bioinformatics, Odisha University of Agriculture and Technology, Bhubaneswar, Odisha, India

B. K. Behera (ed.), *Current Trends in Fisheries Biotechnology*, https://doi.org/10.1007/978-981-97-3165-7_9

9.1 Introduction

Genomic technologies are the most potent tools that have fundamentally altered the field of biological research. In addition, because of the high complexity of genomic variables and the inapplicability of the traditional regression framework, large-scale genetic data provide significant challenges for statistical and bioinformatic data analysis. High-throughput genomic technologies like microRNA arrays, whole genome sequencing, and single nucleotide polymorphisms (SNPs) make it increasingly important to understand how genes work (Chen and Ishwaran 2012).

The massively parallel sequencing of many tiny deoxyribonucleic acid sequences is made possible by next-generation sequencing (NGS) technology. They are particularly well suited for genomics function investigations because of their high throughput, where the main objective is to genomic regions annotation with their physiological functions. Numerous experimental techniques and bioinformatic workflows have been created to make use of NGS's high-throughput capabilities (Guigo and de Hoon 2018).

NGS is a game-changing technology that is reshaping the field of human molecular genetic testing. This enables unparalleled concurrency of sequencing techniques, allowing for tremendously diverse screening methodologies with relatively short turnaround times and reduced budgets. The increasing number of diagnostic laboratories are adopting NGS and using it to develop innovative DNA test solutions ranging in size from multigene disease-specific modules to entire exomes, as well as the rapidly expanding use of whole-genome sequence analysis (Designed Research et al. 1909; Laduca et al. 2014; Rehm 2013; Yang et al. 2013). Additionally, NGS technologies for RNA sequencing, epigenetic profiling, chromatin I immunoprecipitation sequencing, using methylation, and microbial and microbiome sequencing give up new avenues for clinical assessment (Armstrong et al. 2019; Florian Fricke and Rasko 2014; Oliver et al. 2015; Ozsolak and Milos 2011).

9.2 What Is NGS Technology

As next-generation sequencing platforms become more prevalent, there is an increasing need for bioinformatic tools and statistical methods to analyze and manage the enormous amounts of data that these platforms generate. It is used to sequence DNA and RNA and find variants and mutations. An entire genome or tens of thousands of genes may be sequenced quickly using NGS (Qin 2019). NGS data analysis software is currently prevalent; however, many of these applications remain in the initial phases of commercial availability. A few of the techniques employed include de-novo assembly from genome browsing, base-calling, paired or unpaired reads, alignment of sequence reads to a reference, polymorphism identification, and structural variation discovery (Magi et al. 2010).

Deep sequencing techniques like NGS provide excellent throughput, scalability, and speed just after the sanger sequence. With this method, the nucleotide sequence of whole genomes as well as individual regions of DNA or RNA may be determined. In comparison to the Sanger technique, the NGS technologies provide highly parallel analysis, tremendously high throughput from many samples, and far lower costs. One way they differ from the Sanger method is in this (Kulski 2016).

By using NGS techniques, this is now able to efficiently analyze a large number of samples with minimal financial outlay. NGS technology is superior to first-generation sequencing technologies like Sanger sequencing because, in a single run, it can analyze millions to billions of reads in parallel and it only takes a few hours or days to generate gigabase-sized reads. Every human cell has 23 chromosomes, which are made up of millions of DNA bases. The human genome, for instance, is made up of DNA macromolecules ranging in size from 33–247 mbps and having a total length of 3 billion bps. Relatively, 454 genome sequencer-based NGS sequencing of the human genome took over 15 years to complete, cost around $100 million, and included the cooperation of several laboratories from across the world. However, NGS can only sequence extremely brief DNA segments, which generate millions of reads; it cannot sequence the whole genome at once. This continues to be a problem, particularly for genome assembly systems that need a lot of computing power. The development of NGS technology has increased the number of sequencers produced. NGS technology is nevertheless divided into two types in the literature. We distinguish between the most recent sequencing technologies developed in the NGS setting from the second and third generation sequencing technologies. These techniques, in contrast to the second generation, require the development of amplified sequencing banks before amplified DNA clones may be sequenced (Kchouk et al. 2017; Kulski 2016; Schatz et al. 2010).

Millions of tiny DNA fragments are concurrently sequenced by each NGS device. By comparing each read to the human reference genome, bioinformatics techniques are utilized to put these pieces together. Large DNA libraries are created via DNA barcoding, which uses standardized species-specific genomic areas (DNA barcodes), mostly for the purpose of identifying unknown specimens (Shokralla et al. 2012). The three billion nucleotides that make up the human genome have undergone many iterations of sequencing. This provides sufficient depth to provide precise information and unexpected insight into DNA variety.

9.3 Fundamental Aspects of the NGS

First-generation genomic technologies significantly increased the capacity for data collection. Second-generation technologies like whole genome sequencing and tiling microarrays have the potential to significantly enhance data collection. Here, chromatin immunoprecipitation data collected by high-throughput sequencing and tiling arrays are being processed in a time-consuming downstream manner. Gene Track is a new piece of software that has been designed to automate and streamline this process. Smoothing, peak detection, fitting, and visualization are just a few of the components of a typical data processing process that this application unifies into a single workflow. It also works as intended on hardware with a limited feature set (Albert et al. 2008). It will be shown that the majority of so-called "massively parallel" or "next-generation" sequencing techniques and tools nevertheless have close conceptual connections to Sanger sequencing in terms of its basic enzymological foundations. Instead of not separating sequence ladder separation and data collection from enzymatic nucleotide incorporation, massively parallel sequencing techniques differ from Sanger sequencing. As opposed to this, NGS technology executes enzymology and data gathering in an organized and sequential manner, allowing the simultaneous creation of sequence data from tens of thousands to billions of templates. Because of this improved data-generation capability, the phrase "massively parallel" is used to describe DNA sequencing and its applications (McCombie et al. 2019).

9.4 Genome Sequence and Annotation

In the current research, shotgun sequencing is the preferred method for genome sequencing. The first phase is the fragmentation of the genome. Up to a certain limit set by the technology in use, each is sequenced independently. The sequence readings are reassembled into larger, more continuous sequence stretches using cutting-edge computational methods in de-novo assembly (contigs). Sequence read overlap is important for correct assembly at every point in the genome, so high sequencing coverage (also called "read depth") is needed. Due to overlap, longer sequence reads could need a shorter raw read. To provide read position information for the assembly, larger segments are sequenced from both ends. Contigs are often linked together after the first assembly to produce bigger sequencing sections (known as scaffolds). To achieve this, vast DNA fragment libraries that include a large number of kilobases (kb) of genomic sequence are made, and the ends of those segments are sequenced. These libraries are known as paired-end, jump, or mate-pair libraries, depending on the technology and preparation details. If two separate contigs contain the terminal sequences of many independent fragments, they are connected to form a scaffold. The gap created by the estimated fragment length of the library, which accounts for the actual separation of the two contigs, is loaded with the harmless base-pair character "N." The investigation of the least understood mechanisms underlying inbreeding depression (directional dominance vs. overdominance, epistasis, many vs. few loci), as well as the assessment of the impact of genetic mutations on gene expression profiles and polycarbonate responses to environmental change, are also made possible by genome-wide analyses. Other applications of genomic technologies in conservation biology include describing environmental samples and determining how environmental disturbances affect microbial populations (Mardis 2008). Nevertheless, the best resource for genomic approaches is a well-annotated and complete genetic sequence. Whole-genome resequencing data with location information along a genome sequence provide the most comprehensive account of individual genomic discrepancies, including structural rearrangements, single nucleotide polymorphisms, copy number variation, sequence repeats, and insertion-deletion. It is anticipated that this data will soon set the standard for genetic studies of natural populations (Ellegren et al. 2012).

A genomic sequence must be annotated with physiologically relevant information to be fully used. Gene models, gene ontology (GO) terminology, functional information, "Kyoto Encyclopedia of Genes and Genomes" (KEGG) pathways, microRNA, and epigenetic alterations may be included (Kanehisa and Goto 2000). Annotation is mainly limited to the protein-coding sequence (CDS) or transcripts in non-model organisms. Automated gene annotation is conceivable for individual research groups in freshly sequenced species with few gene models, but it is challenging. Still, genomic annotation involves bioinformatics and a lot of research. For a genome sequence to be fully used, physiologically relevant information must be added to it. This data may include microRNA and epigenetic changes, as well as gene models and functional features like KEGG pathways and GO concepts. Annotation is frequently confined to transcripts or, more generally, protein-coding sequences in the area of genetically non-model animals. Although newly sequenced organisms' genes are extremely difficult to annotate genes in where pre-existing models of genes are frequently unavailable, automated gene annotation has theoretically improved to the point where it is now feasible for individual research groups. However, completing a genome annotation requires bioinformatic competence and is a substantial undertaking. The accuracy of the genome assembly has a significant impact on effective annotation. The only genomes that can produce satisfactory results are those that are continuous and nearly full (90%). Longer genomes typically include longer genes, necessitating the use of more contiguous assemblies for successful annotation. The annotation procedures could be conceptually split into two stages. The first stage is the "computational phase," during which various lines of evidence from species-specific transcriptome information or other chromosomes are employed concurrently to provide initial gene and transcript predictions; The second stage is the "predictive phase," in which numerous lines of evidence from species-specific transcriptome data or other genomes are utilized to make predictions. In a step of second annotation, all data is generated into a gene annotation based on a set of criteria given by the annotation pipeline. Masking repeated sequences, including transposable elements and low-complexity areas, is crucial prior to gene prediction. It is important to develop a species-specific repeat repository using programs such as Repeat Mo-dealer or Repeat Explorer since repetitions are frequently poorly preserved across species (Novák et al. 2013; Ekblom and Wolf 2014).

9.5 Comparative Genomics

To decipher biological sequences and draw useful conclusions, comparative genomes play a crucial role. In comparative genomics, proteome comparisons between two or more species are a major aspect. It also involves contrasting features such as gene sequence, gene placement, and regulatory mechanisms. Genealogical events, including deletions, duplications, and horizontal gene transfer, are also studied. The purpose of these kinds of studies is to develop models and principles that can explain such occurrences beyond just describing their similarities and differences (Tatusov et al. 2000). Gene function prediction is a primary motivation for comparative genomics experiments. Even in well-researched species like *Escherichia coli* (4600 genes) and *Saccharomyces cerevisiae* (6500 genes), about 60–70% of the genes have known or projected functions. One key objective is to discover how the remaining 30–40% of genes function. A new set of resources, vocabularies, and specialized lexicons have been developed to facilitate the study of comparative genomics. Important definitions of many key concepts are provided here. When two characteristics have a common ancestor and then diverge, like in the case of two proteins with similar sequences, we say that they are homologous. Homologs are pairings of genes or protein sequences shared by two species that have a common ancestry. Homologs and orthologs, paralogs, and even xenologs may share characteristics. Orthologs are homologs that share a same ancestor gene. Majority of the time, they provide similar functions. Genes or proteins that are related or created in the same way but have diverged over time owing to genome duplication. They often fulfill many functions. Xenologs are homologs in which genetic material has moved horizontally across species. Various responsibilities are shared by xenologs and their hosts. If two genes or proteins have similar functions but are not homologs of each other, then they are analogs. Although they provide different functions, they are analogous. To better understand the function, comparative genomics has shown to be an excellent tool for comparing sequences, rearranging gene order, and analyzing regulatory networks. Analyses of this kind may teach scientists more about the recruitment of enzymes into a network. To further understand what drives the selective loss of enzymes in certain lineages and the recruitment of enzymes and domains, specialized tools may be used. What this means is that we may use a technique called comparative genomics. The major part of evolutionary biology is the study of genetic diversity and speciation, and the definition and history of comparative genomics suggest that this field might be useful in this regard. There are a variety of tools and databases available for the genomic analysis which are listed in Table 9.1.

Pathophysiological knowledge about other species may be gained, and the identification of disease-causing genes in humans can be aided by studying comparative genomics. The finding and creation of novel pharmaceutical targets is another major benefit of such studies (Anon n.d.). Genes of this kind might include virulence genes, unseen but essential genes, or genes unique to a certain species (Anon n.d.).

9.6 Functional Genomics

The identification and cataloging of functional areas in genomes have also been extensively done using comparative genomics techniques. The basic premise is that functional genomic areas have undergone higher evolutionary conservation than non-functional genomic regions. In particular, protein-coding genes in freshly sequenced genomes are annotated using genome sequence comparisons. Less than 10 percent of lncRNAs are conserved across the two species, in contrast to the 80% of human protein-coding genes that can be found in mice (Marques and Ponting

Table 9.1 Tools for genomic data analysis

Databases	Function	URL
AssociationDB	Explore genome-wide association studies	http://genetik.charite.de/AssociationDB
CAMERA	Find information about microbial biological data	http://camera.calit2.net/
CARGO	An open platform system that aims to make it easier to analyze biological information by helping with things like visualizing, mapping, and getting information from the literature	*http://cargo.bioinfo.cnio.es*
CEBS	Public repositories of toxicogenomics data	http://cebs.niehs.nih.gov
CORG	Define homologous non-coding areas in the upstream sections of potential orthologous gene pairs from humans, mice, rattus, tetraodontidae, house mouse, zebra danio	http://corg.molgen.mpg.de/
CG	Conduct relative biometric analysis of chromosomes of different organisms	http://www2.unil.ch/comparativegenometrics/
DEG	Understand about genes that are required for life in both eukaryotes and prokaryotes	http://tubic.tju.edu.cn/deg/
DNAlive	A program for analyzing and visualizing the structural and physical properties of genomic DNA	http://mmb.irbbarcelona.org/DNAlive/
EGO	Study eukaryotic organisms for homologous genes	http://www.tigr.org/tdb/tgi/ego/
EGassembler	Alignment and assemble sequence segments generated by shotgun sequencing or gene transcripts (EST) fragments to reconstitute the source section or gene	http://egassembler.hgc.jp/
EPGD	Find eukaryotic paralog/paralogon information	http://epgd.biosino.org/EPGD/
Endeavor	Web-based platform for prioritizing genes across various species	http://www.esat.kuleuven.be/endeavourweb
Entrez genome	Search for genomic sequences from completely sequenced organisms and those that are being sequenced	http://www.ncbi.nlm.nih.gov/entrez/query.fcgi?db=Genome
EuMicroSatdb	A database of molecular markers found in eukaryotic genomes that have been sequenced	http://ipu.ac.in/usbt/EuMicroSatdb.htm
FGF	Discover gene families and create phylogenetic trees	http://fgf.genomics.org.cn/
FusionDB	Analyze prokaryotic gene fusion events	http://igs-server.cnrs-mrs.fr/FusionDB/

(continued)

Table 9.1 (continued)

Databases	Function	URL
GGT	The ability to view and analyze genetic data is the primary goal of software of this kind	http://www.plantbreeding.wur.nl
GMDD	Methods for detecting genetically altered organisms are cataloged in a database	http://gmdd.shgmo.org/
GWAsimulator	Permits simulation of effects that are not multiplicative and 2 different interaction effects	http://biostat.mc.vanderbilt.edu/twiki/bin/view/Main/GWAsimulator
GIB-V	Search this extensive database for details on viral genomes and segments	http://gib-v.genes.nig.ac.jp/
Genomes OnLine database (GOLD)	Access information regarding complete and ongoing genome and metagenome projects around the world	http://www.genomesonline.org/
HGT-DB	Find potential horizontally transmitted genes by exploring prokaryotic genomes	http://www.fut.es/~debb/HGT
IMG	Explore and evaluate the genomes of microorganisms using a comprehensive database provided by the joint genome institute of the DOE (JGI)	http://img.jgi.doe.gov/
IMG/M	A data management and analysis system for metagenomes	http://img.jgi.doe.gov/m
MBGD	Conduct comparative analysis of completely sequenced microbial genomes	http://mbgd.genome.ad.jp/
ODB	Investigate the bacterial genomes in search of knowledge on operons	http://odb.kuicr.kyoto-u.ac.jp/
OriDB	Look for *S. Cerevisiae* DNA replication initiation locations that are both verified and expected	http://www.oridb.org
OrthoDB	Find groups of orthologous genes	http://cegg.unige.ch/orthodb
PEDANT	Perform a functional and structural examination of the whole genome	http://pedant.gsf.de/
PGMapper	A software program that can automatically match phenotypes to genes from a certain part of the genome or a set of genes	http://www.genediscovery.org/pgmapper/index.jsp
dbGap	Look for information that comes from broad association (GWA) research	http://www.ncbi.nlm.nih.gov/entrez/query.fcgi?db=gap
PSI SGKB	Developed with the purpose of transforming the findings of the protein structure initiative into information that may contribute to a better understanding of biological systems and illness	http://kb.psi-structuralgenomics.org/
PhylomeDB	Database of complete phylomes derived for different genomes within a specific taxonomic range	http://phylomedb.org/

2009; Pervouchine et al. 2015). The degree of interspecies conservation of gene expression patterns and regulatory networks is further clarified by comparative transcriptome studies (Gerstein et al. 2014; Merkin et al. 2012). Due to the mouse's key role as a model of human biology, the similarities between humans and mice have gotten a lot of attention (Breschi et al. 2017; Yue et al. 2014). Although there is a strong correlation between human and mouse gene expression levels averaged across RNA samples, significant variances are discovered when comparing the transcriptome of comparable types of cell (Jubb et al. 2016; Pishesha et al. 2014). Overall, it seems that whether or not the patterns of expression are the same between humans and mice largely depends on the gene. Some genes' expression differs a lot between species but not much between organs, while other genes' expression differs more between organs than between species (Guigo and de Hoon 2018).

9.7 Pipeline for NGS Data Analysis

Raw data must be pre-processed (demultiplexing, quality checking, and adaptor cutting) before non-trivial information can be extracted, such as the detection of active or deactivated signaling networks or the identification of clinically important alterations. Because the NGS principle creates a huge and diverse data collection, genomics and bioinformatics are increasingly considered sciences that need big data methodologies. While advances in technology have made NGS research more affordable and accessible, a current obstacle is the sheer volume of data that must be processed. Complete genome or transcriptome sequencing generates terabytes of raw data, even for a small effort with just 10–30 samples (FASTQ files). To match this data to a reference genome, the sequencing coverage and read length are employed. As a result, the NGS platform creates a large amount of data, making genomics and bioinformatics disciplines that need Big Data approaches (Fig. 9.1). Despite the fact that technological developments have made NGS research more accessible in terms of availability and price, the current restriction is the capability to analyze the massive amount of data generated (Canzoneri et al. n.d.).

9.8 De-Novo Genome Assembly

The term "*de-novo* genome assembly" refers to the process of reconstructing sequence segments chosen at random to figure out the order of each base in a genome. Sequence fragments are patched together to form contigs, and ideally, there would be just one contig per chromosome. Using short-read technology, the de-novo assembly of mammalian genomes has been a challenge owing to the inability of short reads to cover repetitive areas, leading to hundreds of thousands of gaps. A large body of research shows that long-read genome assemblies are far more consistent than those obtained using Sanger sequencing or earlier short-read methods. However, despite the decreasing cost of sequencing, de-novo assembly of reads into high-quality genomes is still complicated and time-consuming x (Bradnam

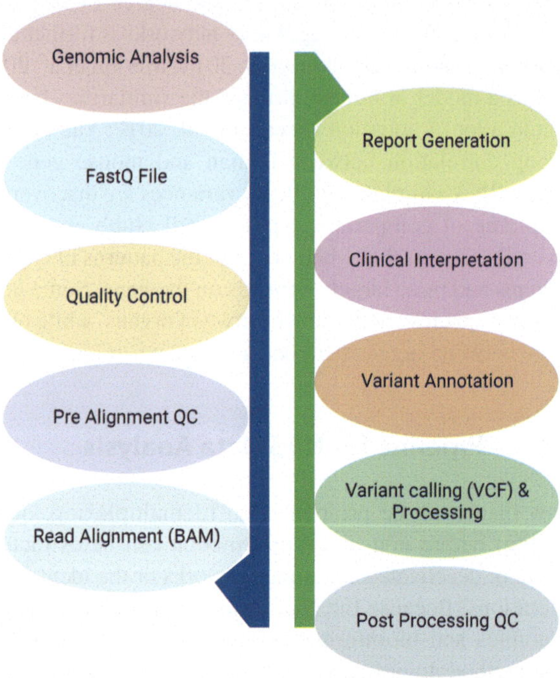

Fig. 9.1 Pipelines of genomic data analysis

et al. 2013; Schatz et al. 2010). Complex de-novo assembly computer approaches attempt to assemble reads into longer, continuous conserved sections by using linkage information from mate-pair reads to construct greater scaffolds (Fig. 9.2). However, the majority of the reads produced are too short, include errors, and are not uniformly spread throughout the genome. Furthermore, genomes might include a lot of repeated elements, that are difficult to put together and frequently lead to mistakes that lower the quality of future polymorphism assessments (Bradnam et al. 2013; Briskine and Shimizu 2017; Treangen and Salzberg 2012). Heterozygosity is a common issue in many diploids and polyploid species, leading to complications during assembly (You et al. 2013). Many heterozygous areas are broken up into smaller pieces called contigs (Pryszcz et al. 2016). Because of this, genomic assemblies might produce fragmented contigs or scaffolds. Incorrectly assembled or erroneous sequence segments may be present in these contigs or scaffolds (Ekblom and Wolf 2014). In recent research, researchers have been using the utilization of larger reads to eliminate cyclical content and create larger scaffolds (Lischer and Shimizu 2017).

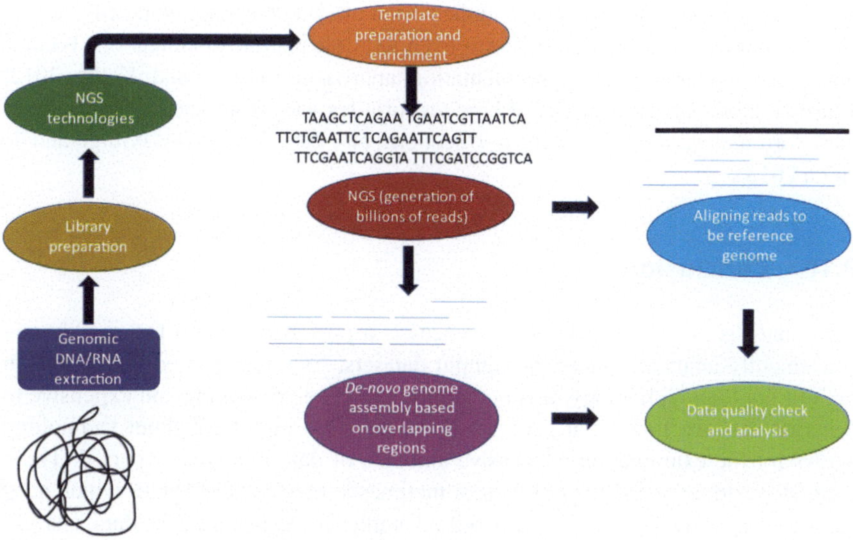

Fig. 9.2 Workflows for next-generation sequencing analysis

9.9 Genomic Protocols

Open euchromatin and closed heterochromatin are two types of chromatins. Transposases generate DNA fragments from accessible chromatin, which are then organized to detect open genomic areas in a certain type of cell or circumstance (Buenrostro et al. 2013). A small number of cells may produce excessive digestion, resulting in reads mapping to closed chromatin regions, whereas a large number of cells may produce inadequate digestion, resulting in longer segments that appear difficult to sequence (Buenrostro et al. 2015). Methylation status can be evaluated by endonuclease digestion of DNA, bisulfite treatment to convert unmethylated cytosine to uracil or immunological precipitation of methylated chromatin followed by sequencing (ChIP-seq). Cytosines (unmethylated) that don't change into uracil may be misinterpreted as methylated. Limiting DNA degradation during bisulfite treatment prevents DNA fragmentation, which hinders PCR amplification. Teta-assisted bisulfite sequencing may be able to resolve the discrepancies in 5-methylcytosine and 5-hydroxymethylcytosine present in mammalian genomes, but those differences cannot be distinguished by traditional bisulfite sequencing (Kriaucionis and Heintz n.d.; Tahiliani et al. 2009; Yu et al. 2012). ChIP-seq-based methods suffer from genomic variations in CpG density and cannot measure DNA methylation with single-nucleotide accuracy. ChIP-seq can also be employed to examine the tendency of histone modifications or transcription factors to bind to genomic areas, depending on the availability of an appropriate antibody (Mikkelsen et al. 2007; Valouev et al. 2008). To achieve greater resolution while using fewer cells a sensitive but focused antibody is required for accurate CHIP-seq data (Furey 2012; Lara-Astiaso et al. 2014). A sensitive but focused antibody is required for accurate ChIP-seq data. Genomic studies often offer a "epigenomic profile" of the entire genome. For instance, ChIP-seq from multiple histone modifications is combined and integrated using segmentation approaches (Ernst and Kellis 2012; Hoffman et al. 2012). In terms of how test profiles vary from one test to the next, these formulas divide the genome into sections, also called fragments (Guigo and de Hoon 2018).

9.10 Conclusion

Data analysis based on genome sequences is a powerful method for quickly and conveniently analyzing large throughput datasets. As a result of advancements in high-throughput technology, it is now much less time-consuming and expensive to generate massive functional genomic data sets. This high-throughput technology allows for the examination of massive amounts of data in a shorter period. These methods yield large amounts of data, sometimes known as "Big Data," which need the use of analytical tools as well as skilled employees to integrate the data, analyze it, and then communicate it to doctors. This topic explains the ideas of customized Big Data, as well as the most significant advancements in genomics and bioinformatics, as well as the opportunities and challenges that lie ahead for these fields. Genetic data will be learned, analyzed, and interpreted with the use of open-source programs like R and Bioconductor. Maintaining a connection between functional genomic data and visible phenotypes is an ongoing challenge.

References

Albert I, Wachi S, Jiang C, Franklin Pugh B (2008) GeneTrack—a genomic data processing and visualization framework. Bioinformatics 24(10):1305–1306

Anon (n.d.) Comparative genomics definition and development of comparative genomics

Armstrong J, Fiddes IT, Diekhans M, Paten B (2019) Whole-genome alignment and comparative annotation. Ann Rev Anim Biosci 7:41–64

Bradnam KR, Fass JN, Alexandrov A, Baranay P, Bechner M, Birol I, Boisvert S, Chapman JA, Chapuis G, Chikhi R (2013) Assemblathon 2: evaluating de novo methods of genome assembly in three vertebrate species. GigaScience 2(1):10

Breschi A, Gingeras TR, Guigó R (2017) Comparative transcriptomics in human and mouse. Nat Rev Genet 18(7):425–440

Briskine RV, Shimizu KK (2017) Positional bias in variant calls against draft reference assemblies. BMC Genomics 18(1):263

Buenrostro JD, Giresi PG, Zaba LC, Chang HY, Greenleaf WJ (2013) Transposition of native chromatin for fast and sensitive Epigenomic profiling of open chromatin, DNA-binding proteins and nucleosome position. Nat Methods 10(12):1213–1218

Buenrostro JD, Beijing W, Chang HY, Greenleaf WJ (2015) ATAC-Seq: a method for assaying chromatin accessibility genome-wide. Curr Protoc Mol Biol 2015:21.29.1–21.29.9

Canzoneri R, Lacunza E, and Abba MC (n.d.) Big data in biomedicine 587 Genomics and bioinformatics as pillars of precision medicine in oncology

Chen X, Ishwaran H (2012) Random forests for genomic data analysis. Genomics 99(6):323–329

Designed Research, MRPLC, ABSO (1909) Genetic diagnosis by whole exome capture and massively parallel DNA sequencing. Proc Natl Acad Sci U S A 106:19096
Ekblom R, Wolf JBW (2014) A field guide to whole-genome sequencing, assembly and annotation. Evol Appl 7(9):1026–1042
Ellegren H, Smeds L, Burri R, Olason PI, Backström N, Kawakami T, Künstner A, Mäkinen H, Nadachowska-Brzyska K, Qvarnström A, Uebbing S, Wolf JBW (2012) The genomic landscape of species divergence in Ficedula flycatchers. Nature 491(7426):756–760
Ernst J, Kellis M (2012) ChromHMM: automating chromatin-state discovery and characterization. Nat Methods 9(3):215–216
Florian Fricke W, Rasko DA (2014) Bacterial genome sequencing in the clinic: Bioinformatic challenges and solutions. Nat Rev Genet 15(1):49–55
Furey TS (2012) ChIP-Seq and beyond: new and improved methodologies to detect and characterize protein-DNA interactions. Nat Rev Genet 13(12):840–852
Gerstein MB, Rozowsky J, Yan KK, Wang D, Cheng C, Brown JB, Davis CA, Hillier L, Sisu C, Li JJ (2014) Comparative analysis of the transcriptome across distant. Nature 512(7515):445
Guigo R, de Hoon M (2018) Recent advances in functional genome analysis [version 1; referees: 2 approved]. F1000Research 7:F1000 Faculty Rev-1968
Hoffman MM, Buske OJ, Wang J, Weng Z, Bilmes JA, Noble WS (2012) Unsupervised pattern discovery in human chromatin structure through genomic segmentation. Nat Methods 9(5):473–476
Jubb AW, Young RS, Hume DA, Bickmore WA (2016) Enhancer turnover is associated with a divergent transcriptional response to glucocorticoid in mouse and human macrophages. J Immunol 196(2):813
Kanehisa M, Goto S (2000) KEGG: Kyoto encyclopedia of genes and genomes. Nucleic Acid Res 28:27
Kchouk M, Gibrat JF, Elloumi M (2017) Generations of sequencing technologies: from first to next generation. Biol Med 9(03):1
Kriaucionis S and Heintz N (n.d.) The nuclear DNA base, 5-hydroxymethylcytosine is present in brain and enriched in Purkinje neurons
Kulski JK (2016) Next-generation sequencing—an overview of the history, tools, and 'Omic' applications. In: Next generation sequencing—advances, applications and challenges. InTech, London
Laduca H, Stuenkel AJ, Dolinsky JS, Keiles S, Tandy S, Pesaran T, Chen E, Gau CL, Palmaer E, Shoaepour K, Shah D, Speare V, Gandomi S, Chao E (2014) Utilization of multigene panels in hereditary cancer predisposition testing: analysis of more than 2,000 patients. Genet Med 16(11):830–837
Lara-Astiaso D, Weiner A, Lorenzo-Vivas E, Zaretsky I, Jaitin DA, David E, Keren-Shaul H, Mildner A, Winter D, Jung S, Friedman N, Amit I (2014) Immunogenetics. chromatin state dynamics during blood formation. Science 345(6199):943–949
Lischer HEL, Shimizu KK (2017) Reference-guided de novo assembly approach improves genome reconstruction for related species. BMC Bioinformatics 18(1):474
Magi A, Benelli M, Gozzini A, Girolami F, Torricelli F, Brandi ML (2010) Bioinformatics for next generation sequencing data. Genes 1(2):294–307
Mardis ER (2008) Next-generation DNA sequencing methods. Annu Rev Genomics Hum Genet 9:387–402
Marques AC, Ponting CP (2009) Catalogues of mammalian long noncoding RNAs: modest conservation and incompleteness. Genome Biol 10(11):1
McCombie WR, McPherson JD, Mardis ER (2019) Next-generation sequencing technologies. Cold Spring Harb Perspect Med 9(11):a036798
Merkin J, Russell C, Chen P, Burge CB (2012) Evolutionary dynamics of gene and isoform regulation in mammalian tissues. Science 338(6114):1593
Mikkelsen TS, Manching K, Jaffe DB, Issac B, Lieberman E, Giannoukos G, Alvarez P, Brockman W, Kim TK, Koche RP (2007) Genome-wide maps of chromatin state in pluripotent and lineage-committed cells. Nature 448(7153):553–560

Novák P, Neumann P, Pech J, Steinhaisl J, MacAs J (2013) RepeatExplorer: a galaxy-based web server for genome-wide characterization of eukaryotic repetitive elements from next-generation sequence reads. Bioinformatics 29(6):792–793

Oliver GR, Hart SN, Klee EW (2015) Bioinformatics for clinical next generation sequencing. Clin Chem 61(1):124–135

Ozsolak F, Milos PM (2011) RNA sequencing: advances, challenges and opportunities. Nat Rev Genet 12(2):87–98

Pervouchine DD, Djebali S, Breschi A, Davis CA, Barja PP, Dobin A, Tanzer A, Lagarde J, Zaleski C, See LH (2015) Enhanced transcriptome maps from multiple mouse tissues reveal evolutionary constraint in gene expression. Nat Commun 6:5903

Pishesha N, Thiru P, Shi J, Eng JC, Sankaran VG, Lodish HF (2014) Transcriptional divergence and conservation of human and mouse erythropoiesis. Proc Natl Acad Sci USA 111(11):4103–4108

Pryszcz LP, Gabaldón T, Gabaldon G (2016) Redundans: an assembly pipeline for highly heterozygous genomes. Nucleic Acids Res 44(12):113

Qin D (2019) Next-generation sequencing and its clinical application. Cancer Biol Med 16(1):4–10

Rehm HL (2013) Disease-targeted sequencing: a cornerstone in the clinic. Nat Rev Genet 14(4):295–300

Schatz MC, Delcher AL, Salzberg SL (2010) Assembly of large genomes using second-generation sequencing. Genome Res 20(9):1165–1173

Shokralla S, Spall JL, Gibson JF, Hajibabaei M (2012) Next-generation sequencing technologies for environmental DNA research. Mol Ecol 21(8):1794–1805

Tahiliani M, Koh KP, Shen Y, Pastor WA, Bandukwala H, Brudno Y, Agarwal S, Iyer LM, Liu DR, Aravind L, Rao A (2009) Conversion of 5-Methylcytosine to 5-Hydroxymethylcytosine in mammalian DNA by MLL partner TET1. Science 324(5929):930–935

Tatusov RL, Galperin MY, Natale DA, Koonin EV (2000) The COG database: a tool for genome-scale analysis of protein functions and evolution. Nucleic Acids Res 28:33

Treangen TJ, Salzberg SL (2012) Repetitive DNA and next-generation sequencing: computational challenges and solutions. Nat Rev Genet 13(1):36–46

Valouev A, Johnson DS, Sundquist A, Medina C, Anton E, Batzoglou S, Myers RM, Sidow A (2008) Genome-wide analysis of transcription factor binding sites based on ChIP-Seq data. Nat Methods 5(9):829–834

Yang Y, Muzny DM, Reid JG, Bainbridge MN, Willis A, Ward PA, Braxton A, Beuten J, Xia F, Niu Z (2013) Clinical whole-exome sequencing for the diagnosis of Mendelian disorders. N Engl J Med 369(16):1502–1511

You M, Yue Z, He W, Yang X, Yang G, Xie M, Zhan D, Baxter SW, Vasseur L, Gurr GM, Douglas CJ (2013) A heterozygous moth genome provides insights into herbivory and detoxification. Nat Genetics 45:220

Yu M, Hon GC, Szulwach KE, Song CX, Jin P, Ren B, He C (2012) Tet-assisted bisulfite sequencing of 5-Hydroxymethylcytosine. Nat Protoc 7(12):2159–2170

Yue F, Cheng Y, Breschi A, Vierstra J, Weisheng W, Ryba T, Sandstrom R, Ma Z, Davis C, Pope BD, Shen Y (2014) A comparative encyclopedia of DNA elements in the mouse genome. Nature 515(7527):355

Soil Biotechnology: A Promising Approach for Wastewater Treatment

10

Tanushree Banerjee, Dhruba Jyoti Sarkar, Vikash Kumar, and Manoharmayum Shaya Devi

Abstract

Water is among the most precious resources in the world, yet due to climate change, rapid population increase, and improper waste management, it is constantly in danger. Restoration and regeneration of the effluent in this situation would aid in better utilising the freshwater resources that are already accessible. Recently there are several green technologies reported to reclaim wastewater with good water quality parameters without much contributing to global warming. One such green technology involving soil biotechnology (SBT) was reported to achieve better treatment of municipal wastewater and industrial effluent without the disadvantages of conventional treatment technologies. SBT for wastewater treatment uses natural resources like earth, pebbles, and sand as well as living media like earthworms, microbes, and flora to treat wastewater in an efficient, less expensive, and environmentally friendly manner. The present chapter focuses on what is SBT, structure of SBT, and how soil biotechnology processes overcome the drawbacks of conventional water treatment plants.

Keywords

Wastewater · Soil biotechnology · Green technology · Earth worm

T. Banerjee (✉) · D. J. Sarkar · V. Kumar · M. S. Devi
Aquatic Environmental Biotechnology and Nanotechnology Division, ICAR-Central Inland Fisheries Research Institute, Barrackpore, Kolkata, West Bengal, India

B. K. Behera (ed.), *Current Trends in Fisheries Biotechnology*,
https://doi.org/10.1007/978-981-97-3165-7_10

10.1 Introduction

On our earth, water is a depleted natural resource. Water scarcity and pollution continue to be significant problems in both urban and rural areas of many developed and developing nations. With growth in global population, especially in nations where urbanisation is occurring more quickly, the problem is projected to rise (Boretti and Rosa 2019). The issue of water pollution has become more widespread due to growing industrialisation (Rahman et al. 2021) that releases wastewater as a by-product after using enormous amounts of potable water. Along with population growth, additional causes that add to the dearth of water in many nations across the globe include climate change, urbanisation, the expansion of agricultural activities, industrial activities, and inadequate wastewater management techniques (Flörke et al. 2018). In addition, due to population growth and the diversification of human activities, the output of solid and liquid wastes has drastically expanded in recent years (Hodaifa et al. 2019). Wastewater is build-up of 99.9% water, but the remaining 0.1% is the problem. This 0.1% content of wastewater has the potential to be extremely concentrated in natural aquatic ecosystem, endangering both the environment and public health. Due to lack of proper wastewater treatment lakes, rivers, and other water bodies of developing nations are becoming more contaminated and the situation is becoming more extravagated due to lack of freshwater supply and growing industrial activities (Edokpayi et al. 2017). The main reason for inadequacy of wastewater treatment is primarily due to unfeasible nature for developing nations.

Wastewater includes domestic waste, industrial effluent, agricultural runoff, and water from businesses and organisations, including hospitals. Wastewater may contain chemical, biological, or physical pollutants. As a result, it can be hazardous for human use. Untreated sewage can cause serious illness if it gets into the public water supply. Untreated wastewater discharged into the environment has a negative impact on the environment, human well-being, and economy.

Only a very small portion of the wastewater that is generated by sewage disposal regions is treated in emerging economies. Inadequate wastewater treatment is typically the result of a lack of financing, but it can also be attributed to ignorance of inexpensive treatment options and the financial benefits of recycled wastewater. Water recycling is now a must for a civilisation to function effectively and to meet the unbound needs of the growing population. To serve this purpose, many conventional technologies like ASP (Activated Sludge Process), UASB (Upflow Anaerobic Sludge Blanket process), Physio-Chemical process, etc. are in practice since last many years. But to assure sustainability, it needs significant startup, operating, and maintenance costs as well as trained labour (Kamble et al. 2017). Natural wastewater treatment systems namely Constructed Soil Filters (CSFs) overcome these limitations. Worldwide, wastewater treatment is performed to remove any existing contamination before it is reused or disposed of into the environment.

10.2 Drawbacks of Conventional Water Treatment Plants

Traditional wastewater treatment facilities are rather large outdoor facilities with a lot of potential negatives.

1. The two major obstacles are the sight and smell of the bulky and industrial water management systems.
2. To achieve sustainability, conventional approaches need significant capital, substantial operating and maintenance budgets, as well as experienced labour.
3. While some methods can reduce phosphorus and nitrogen levels, other methods can combat heavy metals. Most traditional approaches are ineffective at reducing combined heavy metals, harmful, phosphorus, nitrogen, etc.
4. Information on suitable methods for effluent treatment and recycling is not widely shared.

10.3 Natural Technologies for Wastewater Treatment

Engineers have looked for innovative, economical, green, and ecologically responsible ways to prevent water pollution as a result of the high cost of existing treatment processes that have created economic pressures. The concept of grey or environmental biotechnology has been developed that focuses on using biological approach to address waste management problems and environmental issues. To get rid of the suspended particles, organic, and inorganic components of the wastewater, a combination of physical processes including sedimentation, infiltration, and biological processes are used. It also is referred to as the "Constructed Soil Filter" (CSF). Most commonly and latest used is soil biotechnology based on CSF.

10.4 Soil Biotechnology (SBT)

To achieve a pollution-free environment and keeping the environmental biotechnology in consideration, the idea of SBT has emerged. SBT is an Indian technology created by Professor H.S. Shankar of the Indian Institute of Technology-Bombay's Department of Chemical Engineering. It is protected by Indian and US patents (Kamble et al. 2017). SBT has received the Siemens award and was chosen by the Ministry of External Affairs, GOI, as one of the four most innovative technologies in India. The technology is used by companies like Municipal Corporation Greater Mumbai (MCGM), Kalyan Dombivli Municipal Corporation (KDMC), etc. (Gupta et al. 2015).

SBT offers environmentally friendly alternatives to the treatment of wastewater and water resources by utilising microorganisms for the detoxification of contaminated water and soil. SBT has been developed for treating wastewater based on a bio-conversion process and trickling filter mechanism. In this process, the purification is achieved naturally by combination of respiration, photosynthesis, and mineral weathering of micro- and macro-organisms that are housed in a media. SBT has been developed as an oxygen-supplying biological engine, which can treat any sort of water, including home, municipal, and industrial water (Vision Earth Care 2013). The natural process of microbe environment relationship and their dynamic interaction with soil, water, and air is utilised in SBT.

Table 10.1 presents the simplified and gross chemistry of the basic natural processes used into SBT. The energy of respiration (Eq. 10.1) is used in the soil

Table 10.1 Simplified chemistry of chemical reactions at work during biofiltration (Pattanaik et al. 2003)

Eq. 10.1	Respiration	$(CH_2ON_xP_yS_zK_y)_n + nO_2 + nH_2O = nCO_2 + 2nH_2O$ + Minerals (N, P, S, K) + Energy
Eq. 10.2	Photosynthesis	$nCO_2 + 2nH_2O$ + mMinerals (N, P, S, K) + Sunlight = $(CH_2ON_xP_yS_zK_y)_n + nO_2 + nH_2O$
Eq. 10.3	Nitrogen fixation	$N_2 + 2H_2O$ + energy = $NH_3 + O_2$(in soil) $N_2 + 2H_2O$ + light = $NH_3 + O_2$(in water)
Eq. 10.4	Acidogeneis	$4C_3H_7O_2NS + 8H_2O = 4CH_3COOH + 4CO_2 + 4NH_3 + 4H_2S + 8H^+ + 8e$-
Eq. 10. 5	Methanogenesis	$8H^+ + 8e- + 3CH_3COOH + CO_2 = 4CH_4 + 3CO_2 + 2H_2O$
Eq. 10.6	Mineral weathering	$CO_2 + H_2O = HCO_3- + H^+$ Primary minerals + $CO_2 + H_2O = M^{+n} + nHCO_3-$ + soil/clay/sand
Eq. 10.7	Nitrification	$NH_3 + CO_2 + 1.5O_2$ = Nitrosomonas + $NO_2- + H_2O + H^+$ $NO_2- + CO_2 + 0.5O_2$ = Nitrobacter + NO_3-
Eq. 10.8	Denitrification	$4NO_3- + 2H_2O$ + energy = $2 N_2 + 5O_2 + 4OH-$

processes at mesophilic temperatures (20–45 °C), where it is used to extract nutrients from the environment, such as nitrogen, as per Eq. 10. 3. Bioindicator plants are used to remove excess metabolites via photosynthesis, as per Eq. 10.2. The chemistry of acidogenesis controls the generation of acidity due to decomposition, as per Eq. 10.4. In SBT, mineral additives that have been specially formulated to control the pH of the environment are used. Eq. 10.6 describes the chemistry of this weathering reaction; M^{+n} stands for the nutrients released from the primary minerals, and soil, sand, and clay are the by-products of this weathering reaction. Nitrogen regulation involves the assimilation of nitrogen (assimilatory nitrate removal), plant absorption as described by Eqs. 10.7 and 10.8, denitrification as described by Eq. 10.8, and so on. The inputs-outputs from the SBT conversion process are quantified by means of these chemical equations (Pattanaik et al. 2003).

10.5 Physical Structure of SBT

Prestressed-concrete, rock building material, and/or earth bunds are all acceptable materials for SBT systems. It includes a tank for raw water, a container for the bioreactor, a tank for treated water, pipelines, and pumps (Fig. 10.1). The raw water tank, a structure resembling a bioreactor, the treated water tank, pumping equipment, and its electrical installations—if necessary, make up the bulk of the SBT's physical structure. The method is designed to treat sewage from homes and businesses that mostly contain organic effluent. Depending on the need and investment potential, treated water of various quality levels, from river discharge quality up to nearly drinking-water standard, can be obtained from an SBT.

10.6 Constituents (Figs. 10.2 and 10.3)

(a) Media: The appropriate size and composition of primary minerals are created from soil.
(b) Culture and catalyst: Geophagus worms (*Eisenia fetida*), nitrifying and denitrifying creatures, bacteria that can break down protein, starch, cellulose, and lignin, and anaerobic bacteria that can produce methane. Ruminant animals' excretions are used to create the bacterial culture.
(c) Additives: Formulated from organic materials with the right mineral compositions and particle sizes to give sites for CO_2 collection.
(d) Plants: Green plants, especially those with tap roots, serve as bioindicators and enhance aesthetics.
(e) Underdrain: Stone fragments in a range of sizes, from fine sand to gravel.

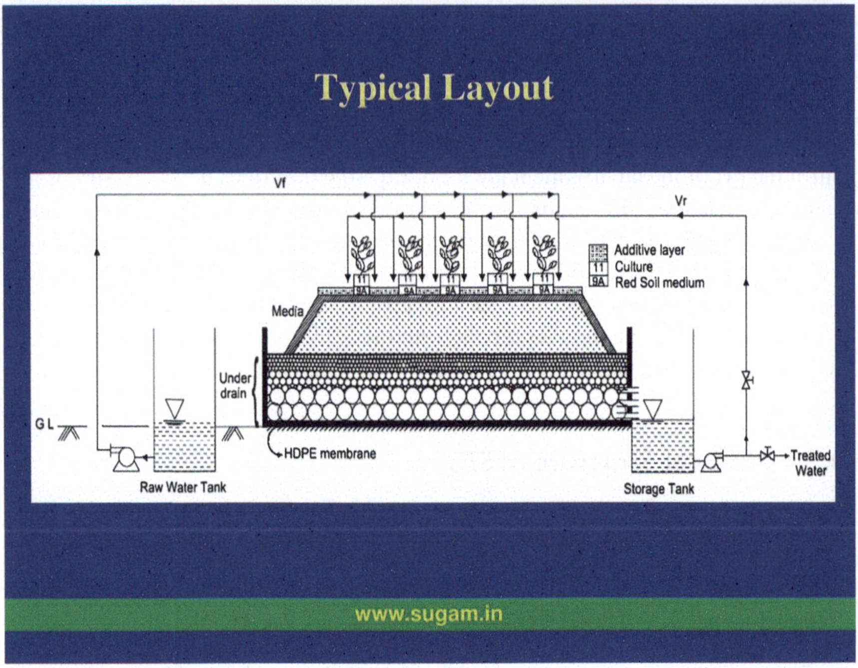

Fig. 10.1 Layout of SBT (Source: Kadam et al. 2008a, b)

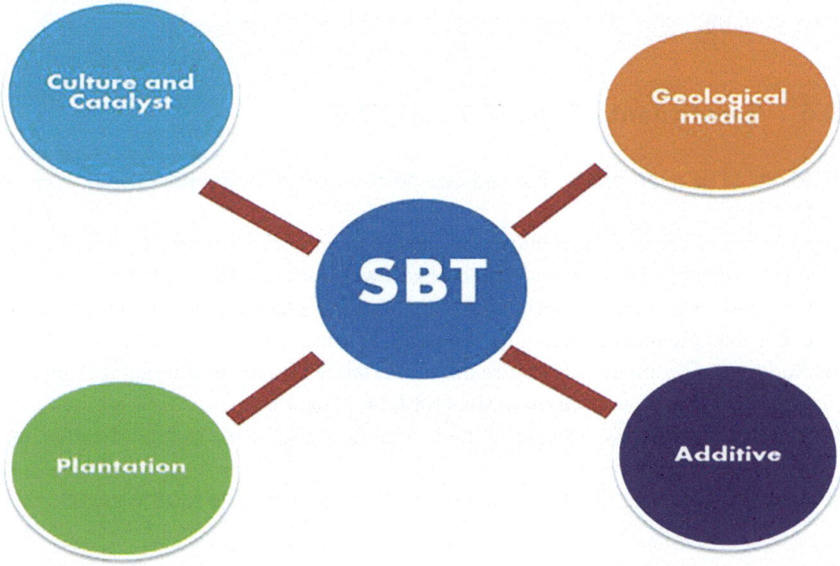

Fig. 10.2 Components of SBT

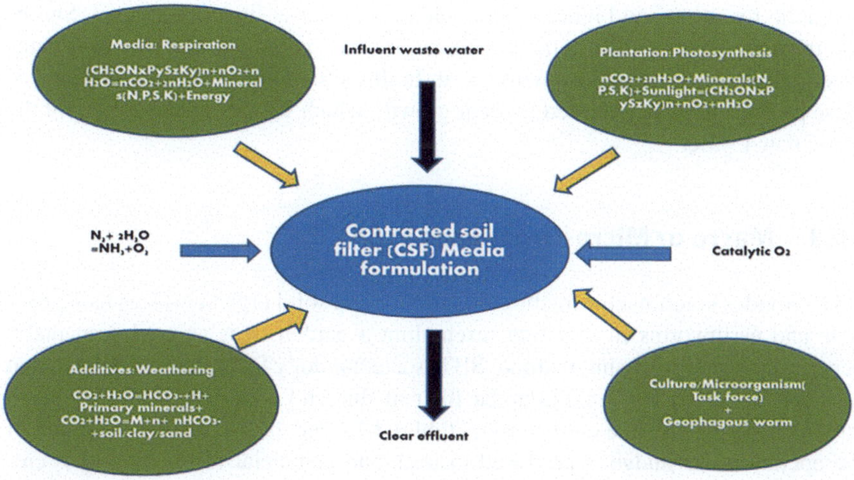

Fig. 10.3 Components of SBT media

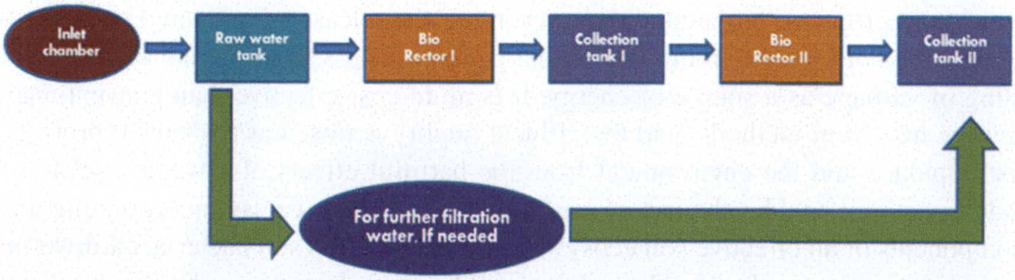

Fig. 10.4 SBT treatment method

10.7 Treatment Method

A holding tank is used to collect the wastewater before it is injected into a bioreactor with a trapezoidal shape. Construction of the bioreactor involves excavation, which is then waterproofed. The base is where the underdrain is placed. The tank is then layered with media and cultural elements. Rows of plants line the bioreactor's exterior. The incoming wastewater is distributed uniformly over the surface of the bioreactor by a surface network of perforated pipes. For aeration, another vertically arranged pair of pipes is also extended into the bioreactor (Fig. 10.4).

Through the network of perforated pipes, water is pumped over the bioreactor and starts to trickle down the filtering media. The top media prevents the wastewater's suspended particles from flowing out. Dissolved contaminants are eliminated as the water flows through the remaining levels, and then treated water exits the tank's bottom outlet and is collected in a storage tank for treated water that has been built next to it. If required, recirculation pumps can be built to return the water to the bioreactor. This results in a second cycle of purification with the appropriate hydraulic retention and higher output quality of water at the desired level. Shrubs and trees are positioned atop the bioreactor to serve as bioindicators, or organisms used to track the state of the environment. In this situation, the ecological health of these plants will be determined by their growth, which will reveal the calibre of the recirculated water.

10.8 Macro or Micro-Organisms in SBT

SBT provides solutions for treating wastewater and solid organic wastes using bacteria and earthworms in a setting resembling a garden. It is an environmentally beneficial waste processing method. SBT is a technology for water purification that uses soil, specially designed granular filter media, and a chosen culture of macro and microorganisms like earthworms (Tables 10.2 and 10.3). At the conclusion of the operation, no sludge is produced instead, and beneficial vermicompost is created. SBT has been found to reduce yearly operation and maintenance costs that are lower than traditional or advanced technologies and comparable to land-based systems. It requires almost no upkeep. It uses the least amount of energy and doesn't create any offensive bio-sludge. By preventing the release of untreated sewage, it safeguards both the environment and the general people. It also enables the recycling of garbage as a source of energy. It is more cost-effective than conventional sewage treatment methods, and the effluent quality is raised as a result. It protects the populace and the environment from the harmful effects of sewage discharge. Additionally, it enables the use of waste as an energy source. By incorporating the components of an effective soil ecosystem, which includes soil bacteria, earthworm and plant species, and mineral fertilisers, SBT aids in harnessing the bioenergy in organic matter. The key benefits of this technology are its cheap maintenance costs, usage of treated effluent, ecological effects, and greater acceptance.

Geophagus earthworms, *Eisenia fetida*, *Pheretima elongate*, and bioindicator plants are a few examples of the micro and macroorganisms found in SBT's artificial ecosystem of formulated media. Organic and inorganic materials are processed

Table 10.2 Specification of the bioreactor media (Pattanaik et al. 2003)

Item	Details
Underdrain	Gravel dp = 25 mm White sand dp = 2 mm
Media	Specific gravity 2.62 BET specific surface area 23 m^2/g CEC = 1.5 g/kg
Soil	Sand: 67%, silt: 23%, clay: 10% Specific gravity: 2.66 BET area: 33.6 m^2/g CEC = 1.5 g/kg
Earthworm	Pheretima elongate

Dp particle diameter, *BET* Brunauer, Emmett & Teller (isotherm)

Table 10.3 Media's physicochemical and microbiological characteristics (Nemade et al. (2009)

Physicochemical properties	
Moisture level (%)	23.4 ± 5.1
Clay (g/kg)	11 ± 1.7
Silt (g/kg)	550 ± 25
Sand (g/kg)	400 ± 20
Gravel (g/kg)	38.0 ± 5.0
Soil texture (USDA scheme)	Sandy loam
Hydraulic conductivity (*KLa*; cm/s)	4.2×10^{-5}
pH soil suspension (1:5)	7.0 ± 0.20
Organic matter (g/kg)	1.3 ± 0.1
Total carbon (g/kg)	4.8 ± 0.4
Cation exchange capacity (m.e./100 g)	41.0 ± 5.8
Anion exchange capacity (m.e./100 g)	1.9 ± 0.4
Microbial properties	
Actinomycetes (CFU/g)	1.2 × 108
Heterotrophic plate count (CFU/g)	2.0 × 1012
Protozoa	
Naked amoebae (cells/g)	8.3 × 106
Flagellates (cells/g)	8.3 × 104
Ciliates (cells/g)	8.3 × 104
Geophagus worm	*Pheretima elongata* present

by bacteria to produce bioconversion, and geophagus worms control the bacterial population. Media, culture, and additives are all present. These are a component of the procedure that oxidises organic waste to clean it while emitting carbon dioxide. The nitrogen burden in wastewater is converted to elemental nitrogen gas through nitrification and then denitrification. A pH buffering effect is produced by the primary minerals that make up the base media in the bioreactors where the purification procedures are carried out. Earthworms are utilised to aerate the environment and manage bacterial populations, while trees and bushes placed on the bioreactor's surface serve as bioindicators to show when the plant is completely functional. COD, BOD, suspended solids, colour, odour, bacteria, and coliforms are all removed in one environmentally friendly facility that is open to the atmosphere. It differs from land treatment, which requires a lot of acreage, and manmade wetlands, which involve aquatic ecology. The groundwater recharge, irrigation, constructions, industrial activities, flushing, gardening, road/car wash, etc. may all be done using the treated water generated by soil biotechnology.

The scientific method of cultivating artificially large generation of earthworms to break down organic food waste into a nutrient-rich material is known as vermiculture. A recently developed method uses earthworms to filter wastewater. The body of an earthworm functions as a biofilter, and it has been discovered that earthworms can remove the 5 days' worth of BOD, COD, total dissolved solids (TDS), and total suspended solids (TSS) from wastewater by ingesting and degrading organic waste, heavy metals, and solids from wastewater as well as by absorbing them through body walls. The earthworms in vermifilter help to increase microbial activity by boosting the number of soil microorganisms. Vermicompost serves as a potent buffer and natural adsorbent for reducing transition metals, biological oxygen demand (BOD), and chemical oxygen demand (COD) from industrial effluents. The earthworms work our soils tirelessly, and their castings are the greatest and richest fertilisers available. They are highly helpful in the sludge and wastewater treatment processes. They are helpful for treating household and residential sewage as well as industrial and domestic wastewater. They remove between 80–90% of BOD and between 70–80% of COD. They transform organic stuff into a stable product after stabilising it. The best and most commonly used earthworm for treating wastewater from various fields is *Eisenia fetida*.

10.9 Life Cycle Assessment (LCA) of SBT

Environmentalists have seen a recent increase in the effectiveness of natural treatment solutions. These technologies are practical because they are simple to maintain, have longer life cycles, and can recover a variety of resources, such as treated effluent for irrigation, organic humus for improving soil, and energy in the form of biogas. LCA is a useful tool for examining environmental effects associated with a good, service, or process from the beginning to the end (Hospido et al. 2004; Gallego et al. 2008; Li et al. 2013). LCA has been frequently employed as a comparative method to assess the variations in conventional systems and environmental systems and to determine whether the former have less detrimental effects (Roux et al. 2010; Pan et al. 2011). The created LCA model aims to assess the plant's environmental effects. Construction can have up to 80% of the influence on natural processes including waterlogged depressions, artificial marshes, waterhole, and time-consuming infiltration systems (Machado et al. 2007). It has been determined that the construction phase has a substantial impact, particularly in terms of global warming potential (GWP) (Doka 2009). Life cycle assessment research should, in accordance with ISO 14040/44, go through the following four stages: aim and scope definition, life cycle inventory, life cycle impact assessment, and result interpretation. Scope and goal clarification: The most crucial and initial part of any LCA is the determination of goals and scope. A measurement of the effectiveness of the product system is the functional unit (FU). It serves as the reference flow in the LCA model to which all other flows are referred (Gallego et al. 2008). System limit: The system limits specify which operations should be part of the LCA and which must be omitted (ISO 14040, 2006). This includes construction, action and upkeep (O&M), as well as freightage and dumping of sludge to landfills, operations related to treated effluent. Because of its low price, low energy requirement, simple functioning, minimal maintenance, low noise, and odour-free features, this technology is frequently regarded as one of the most promising methods for treating wastewater. It also maintained better socioeconomic and environmental benefits when compared to conventional wastewater treatment systems. Additionally, it can be utilised to get rid of heavy metals. The other benefits of this technology include its low maintenance costs, usage of treated effluent, environmental effect, and social acceptance.

10.10 Salient Features of SBT

1. The method can reduce levels of turbidity, nitrogen, phosphorus, total suspended solids (TSS), chemical oxygen demand (COD), and biological oxygen demand (BOD).
2. Tuneable output quality and higher efficiency.

   The food-to-mass ratio (F/M) is one of the most important components of a water treatment facility. F/M ratio is the ratio of the quantity of microorganisms present in the medium being used to treat the water to the quantity of unclean, undesired matter existing in the untreated water. In general, there are two categories of media used to treat wastewater: aquatic (water) treatment and terrestrial (soil) treatment. In comparison to soil, which has 10^{12}–10^{16} cells/g, water has 10^{7}–10^{8} cells/mL of microorganism variety. Therefore, a treatment system based on soil will be more effective than one based in water with a lower F/M ratio due to the increased microbial diversity in soil. Being a terrestrial technology, SBT can effectively treat wastewater and produce results that are superior to any traditional aquatic technology, including activated sludge process (ASP), membrane bioreactor (MBR), single batch reactor (SBR), etc., all while requiring very little operational expense (Table 10.4).
3. No requirement of mechanical Aeration

   SBT differs from previous technologies in that it does not require mechanical aeration, in contrast to ASP. When wastewater is being treated, oxygen is injected to speed up the process of organic matter decomposition. Aeration also stirs the water, allowing the biomass to come into touch with the suspended materials. This aeration method has a disadvantage in that it uses a lot of electrical energy and needs to be maintained for a long time. Additionally, only 21%

Table 10.4 Comparison of different technologies

Items	Soil biotechnology (SBT)	Membrane bioreactor (MBR)	Activated sludge process (ASP)	Single batch reactor (SBR)
Water recovery	98%	85%	80%	85%
Water quality	Excellent	Good	Good	Good
Sludge handling	No	Extra	Extra	Extra
Operation and maintenance	Easy	Skilled	Skilled	Skilled
Ambience	Green	Non-green	Non-green	Green
Energy consumption	Low	High	High	High
Mechanical equipment	No	Yes	Yes	Yes

of the air is made up of oxygen, and the remaining nitrogen-containing air is wasted. Poor water quality occurs when the aeration procedure is improperly carried out. In the case of SBT, aeration is removed by organic means. With a specific set of microorganisms, the three primary natural processes of respiration, photosynthesis, and mineral weathering are united in a single medium. There is no requirement to add oxygen from the outside because of the way the mixture is produced.

4. No odour produced.
5. Due to the high rate of natural oxygen transfer during the process (0.06 kWh/kL sewage), very little energy is used.
6. Wastewater has a very low spatial intensity of 0.8–1.0 sq.m/kL per day.
7. An artificially created, perpetual natural process with only pumps as its moving parts.
8. There is no sludge produced.

   The sludge or backwash water disposal generated using other technologies is a significant issue. For instance, in the case of ASP, a significant amount of sludge collects in the settling tank and must be regularly removed and disposed of in order to prevent the water quality from being impacted. The sludge is released into surrounding sewers, where it eventually contaminates another water supply. Sludge drying is another method of controlling sludge production (Gu et al. 2021). In this, the sludge is dispersed over a huge region and allowed to dry until it is transformed into manure or fertiliser. This can occasionally be dangerous because certain toxins may contaminate the surface or groundwater. Additionally, it takes up a lot of space and emits a bad odour. This calls for SBT's capacity to prevent sludge production, which is another advantageous property. Due to the dense population of microbes in the soil, the full BOD load of the plant is managed, and no by-products or gases, such as hydrogen sulphide, ammonia, carbon disulphide, etc., which are responsible for odours in the area around the plant, are produced. As a result, handling sludge requires little space or effort. In reality, this approach has the extra benefit that, when the topsoil layers are replenished in a timely manner, they work as a green manure and can be applied to agriculture.
9. Removal of ammonia, BOD, COD, suspended particles, colour, odour, and very high bacteria.
10. Practically free of maintenance.

    Some of the existing SBT-based plants in India include the following:

    - 3 MLD Wastewater Treatment in the Corporation of Mumbai, Maharashtra
    - The Airport Authority of India's 120 KLD Plant is located at the airport in Udaipur, Rajasthan.
    - The Lucknow Airport's 140 KLD plant in Uttar Pradesh.
    - Chandigarh's 10 KLD plant at a chemical factory.
    - Mumbai's Bombay Presidency Golf Club in Chembur, India.

10.11 Conclusion and Future Prospect

SBT is one of the environment-friendly wastewater treatment technologies with very high potential in minimal requirement of power. Since the SBT technology is not resource-demanding, like no need of skilled labour, and low operational and maintenance cost it would sustain easily in adverse conditions without any breakdown. The treated water through SBT can be used for irrigation, groundwater recharging, industrial process, gardening, flushing, construction, road/car wash, etc. On the other hand, setting up an SBT plant to treat the wastewater would help to prevent the deterioration of rivers to a great extent.

References

Boretti A, Rosa L (2019) Reassessing the projections of the world water development report. NPJ Clean Water 2(1):1–6

Doka G (2009) Life cycle inventories of waste treatment services. Ecoinvent report No. 13. Swiss Centre for Life Cycle Inventories, Dubendorf

Edokpayi JN, Odiyo JO, Durowoju OS (2017) Impact of wastewater on surface water quality in developing countries: a case study of South Africa. Water Quality 10:66561

Flörke M, Schneider C, McDonald RI (2018) Water competition between cities and agriculture driven by climate change and urban growth. Nat Sustain 1(1):51–58

Gallego A, Hospido A, Moreira MT, Feijoo G (2008) Environmental performance of wastewater treatment plants for small populations. Resour Conserv Recycl 52(6):931–940

Gu Z, Yang J, Liu J, Tao L, Zhang Y, Huang L (2021) Study on sewage sludge drying system with built-in solar drying bed. In: E3S web of conferences, vol 237. EDP Sciences, Les Ulis, p 01041

Gupta R, Aggarwal S, Vishal V, Sahu V (2015) Soil biotechnology to treat Nazafgarh drain water before disposal into river Yamuna. CiVEJ 2(4):1

Hodaifa G, Paladino O, Malvis A, Seyedsalehi M, Neviani M (2019) Green techniques for wastewaters. Interface Sci Technol 30:217–240

Hospido A, Moreira MT, Fernández-Couto M, Feijoo G (2004) Environmental performance of a municipal wastewater treatment plant. Int J Life Cycle Assess 9(4):261–271

Kadam A, Oza G, Nemade P, Dutta S, Shankar H (2008a) Municipal wastewater treatment using novel constructed soil filter system. Chemosphere 71(5):975–981

Kadam AM, Oza GH, Nemade PD, Shankar HS (2008b) Pathogen removal from municipal wastewater in constructed soil filter. Ecol Eng 33(1):37–44

Kamble SJ, Chakravarthy Y, Singh A, Chubilleau C, Starkl M, Bawa I (2017) A soil biotechnology system for wastewater treatment: technical, hygiene, environmental LCA and economic aspects. Environ Sci Pollut Res 24(15):13315–13334

Li Y, Luo X, Huang X, Wang D, Zhang W (2013) Life cycle assessment of a municipal wastewater treatment plant: a case study in Suzhou, China. J Clean Prod 57:221–227

Machado AP, Urbano L, Brito AG, Janknecht P, Salas JJ, Nogueira R (2007) Life cycle assessment of wastewater treatment options for small and decentralized communities. Water Sci Technol 56(3):15

Nemade PD, Kadam AM, Shankar HS (2009) Removal of iron, arsenic and coliform bacteria from water by novel constructed soil filter system. Ecol Eng 35(8):1152–1157

Pan T, Zhu XD, Ye YP (2011) Estimate of life-cycle greenhouse gas emissions from a vertical subsurface flow constructed wetland and conventional wastewater treatment plants: a case study in China. Ecol Eng 37(2):248–254

Pattanaik BR, Bhawalkar US, Kadam A, Shankar HS, Manglani J (2003) Soil biotechnology for waste water treatment and utilization. In: 13th IWA-ASPAC Conference & Exhibition, Cebu City, Philippines, pp 181–188

Rahman MM, Alam K, Velayutham E (2021) Is industrial pollution detrimental to public health? Evidence from the world's most industrialised countries. BMC Public Health 21(1):1–11

Roux P, Boutin C, Risch E, Heduit A (2010) Life cycle environmental assessment (LCA) of sanitation systems including sewerage: case of vertical flow constructed wetlands versus activated sludge. In: 12th IWA International conference on wetland systems for water pollution control, vol 2

"Plankton" as Environmental Biotechnological Tool to Assess Ecological Health Status of Aquatic Ecosystem

11

Pritijyoti Majhi, Pranab Gogoi, Pranaya Parida, and Prajnaritambhara Behera Swain

Abstract

Phytoplankton and zooplankton play an important role in fresh and marine ecosystems because of their significant contribution in aquatic food chain as primary and secondary producers. Nearly half of all primary and oxygen production worldwide is made up of phytoplankton. Phytoplankton also significantly accelerates biogeochemical cycles and nutrient (re)cycling in both terrestrial and aquatic ecosystems. Because they are a rich supply of vitamins, essential fatty acids, carbohydrates, proteins, pigments, and biologically active primary and secondary metabolites, plankton have taken on a great deal of significance in the field of biotechnology in recent years. The plankton community is one of the most promising sources for novel products due to its vast biodiversity. Microalgae biotechnology has recently been able to satisfy the high demands of the food and pharmaceutical industries.

Keywords

Phytoplankton · Zooplankton · Biotechnological tool · Aquatic ecosystem · Ecosystem health

P. Majhi (✉)
Aquatic Environmental Biotechnology & Nanotechnology Division, ICAR-Central Inland Fisheries Research Institute, Barrackpore, Kolkata, West Bengal, India

P. Gogoi · P. Parida · P. B. Swain
Reservoir and Wetland Fisheries Division, ICAR-Central Inland Fisheries Research Institute, Barrackpore, West Bengal, India

B. K. Behera (ed.), *Current Trends in Fisheries Biotechnology*,
https://doi.org/10.1007/978-981-97-3165-7_11

11.1 Introduction

The plankton community is a diverse collection of microscopic creatures that live in freshwater and saltwater environments alike. Plankton is a pelagic creature that is floating in water with restricted movement, frequently less than that of the water it transports. Victor Hensen, a German marine biologist, coined the name "plankton" and conducted a number of expeditions to investigate its distribution, abundance, and composition. Algae live in almost all aquatic and wet terrestrial habitats, sometimes even on top of other plant and animal species (Hoffmann 1989; Round et al. 1990). In aquatic environments, algal communities have a variety of lifecycles, with some occurring infrequently, persisting for a short period of time, and then disappearing.

The two categories of plankton are photosynthetic organisms (phytoplankton) and heterotrophic creatures (zooplankton). Plankton can also be divided according to the methods of collection. Total plankton are those that are not filtered through a net but are concentrated from a specified volume of water that has been taken from the aquatic system. Net planktons are those that pass through a known mesh size of net. Plankton can be classified based on its size: Picoplankton-0.2–2.0 μm (*Synecococcus, Aphanothece, Aphanocapsa*), Nanoplankton-2–20 μm (*Chroococcus, Merismopedia, Scenedesmus, Crucigenia, Cryptomonas, Rhodomonas*), Microplankton-20–200 μm (*Ceratium, Peridinium, Pinnularia, Cymbella, Asterionella, Keratella* sp., *Brachionus* sp.), Macroplankton-200–500 μm (*Anabaena, Nostoc, Platyias* sp., *Scaridium* sp.), Mesoplankton-500–1000 μm (majority of Copepods) and Megaplankton- > 1000 μm (mostly marine).

Plankton can also be categorised as holoplankton, which maintains its planktonic state throughout its life cycle, meroplankton, which only maintains its planktonic state for a portion of its existence, and pseudoplankton (which occupy in upper layer of water column accidentally). Plankton are sometimes known by their habitat; plankton from ponds (heleoplankton), of lakes (limnoplankton), of running water (rheoplankton) and of salt water (haliplankton).

An autotrophic element of the plankton community in a freshwater habitat is phytoplankton. Phytoplanktons are important food organisms for both freshwater aquaculture and mariculture and exist naturally in the aquatic environment. Algae and cyanobacteria are the two groups of phytoplankton. Although these two classes have diverse physical forms, they share the capacity for photosynthesis. No of their taxonomy, phytoplankton can all perform photosynthesis for energy since they all have at least one type of chlorophyll (chlorophyll a) (Table 11.1).

Zooplankton is a collective term for a variety of organisms that float and drift and have a limited ability to move. Most of them are microscopic, monocellular or multicellular, and range in size from a few microns to at least one millimetre. There are distinctions in physical characteristics and taxonomic place in addition to size variations. The zooplankton category contains Protozoa, Rotifera, Cladocera, and Copepoda, which are microscopic planktonic organisms with animal-like traits. The zooplankton group, i.e., Cladocera, Rotifera, and Copepoda account for almost 90% of the entire zooplankton population in freshwater habitats (Datta-Munshi et al.

Table 11.1 Algae classes and their pigments

Group	Division	Class	Pigment
Prokaryotic algae	Cyanophyta (cyanobacteria or blue-green algae)	Cyanophyceae	Chlorophyll *a*; phycocyanobilin; phycoerythrobin; Phycobiliproteins
Eukaryotic algae with chloroplasts surrounded only by the two membranes of the chloroplast envelope	Chlorophyta (green algae) Rhodophyta (red algae)	Prasinophyceae Chlorophyceae Ulvophyceae Charophyceae Rhodophyceae	Chlorophylls *a* and *b*; Chlorophyll *a*; Phycobiliproteins; phycoerythrobin,
Eukaryotic algae with chloroplasts surrounded by one membrane of chloroplast endoplasmic reticulum	Euglenophyta (euglenoids) Dinophyta (=Pyrrophyta) (dinoflagellates)	Euglenophyceae Dinophyceae	Chlorophylls *a* and *b*; Chlorophylls *a, c*1and c2; Peridinin, fucoxanthin
Eukaryotic algae with chloroplasts surrounded by more than two membranes of chloroplast endoplasmic reticulum	Cryptophyta (cryptomonads) Heterokontophyta (heterokonts)	Cryptophyceae Chrysophyceae (golden-brown algae) Bacillariophyceae (diatoms) Raphidophyceae (chloromonads) Xanthophyceae (yellow-green algae) Phaeophyceae (brown algae) Prymnesiophyta (haptophytes)	Chlorophyll *a* and *c*; phycobiliproteins Chlorophyll *a* and *c*; fucoxanthin

2010). They are essential players in the aquatic ecosystem's food chain because they transform phytoplankton's energy into a form that larger animals can use. Zooplanktons are critical components of aquatic food webs and play a significant influence in aquatic production in freshwater habitats.

11.2 Phytoplankton and Its Functions

In an aquatic habitat, phytoplankton forms the basic foundation of the nutrient cycle. In order to keep the biotic and abiotic components of an aquatic ecosystem in optimum balance, they are essential. Phytoplankton is the primary producer in an aquatic ecosystem, and an important part of the aquatic food chain. Therefore, phytoplankton is essential to support a healthy aquatic ecosystem, and used as ecological indicator (Xu et al. 2001). The quality of water is identified in terms of its physical, chemical, and biological characteristics. Algae are environmental cleaners on the one hand, and polluting organisms on the other. The variety of ecological niches is indicated by the phytoplankton diversity with seasonal variation.

11.3 Primary Producer

Phytoplankton are crucial components of freshwater aquatic habitats. Most algae are autotrophs, or primary producers, which means they use sunlight to convert water and CO_2 into a source of food and energy for all aquatic animals and produce oxygen as a byproduct. In freshwater aquatic systems, the amount of primary production by algae can be measured using the amount of fixed carbon per unit area with time (mgC/m^{3}h), which varies from environment to environment. If there are large nutrient loads like nitrogen and phosphorus, there will be noticeable amounts of primary productivity at the surface of eutrophic lakes. Significant levels of primary productivity will be seen at the surface of eutrophic lakes if high nutrient loads like nitrogen and phosphorus are present. In contrast, mesotrophic and oligotrophic lakes have poor overall productivity, but because of better light penetration, productivity extends deeper into the water column.

The primary producers' cyanobacteria, also referred to as blue-green algae, are very important and can be found in a variety of freshwater habitats, including those with high and low nutrient concentrations as well as phytoplanktonic and benthic environments. The water's surface is typically ruled by these algae. These algae can bloom and be a problem in a mesotrophic to eutrophic environment. For instance, in eutrophic conditions, genera of colonial algae, such as *Microcystis*, *Anabaena*, etc., grow rapidly and release the neurotoxin microcystin, which kills aquatic creatures. However, these species can also appear sporadically in small numbers. When nutrient levels are extremely high, a change from colonial blue-green to green algae can occur as a large bloom develops. This type of condition is common in some fish culture ponds as a result of intensive organic and manure treatment to boost fish productivity. In this scenario, an early summer diatom bloom (*Navicula* and *Nitzschia*) is replaced by rapidly spreading unicellular and small colonial green algae, such as *Scenedesmus* and *Pediastrum*. The amount of light and nutrients available determines whether blue-green algae will float or sink; they have the ability to change their buoyancy. All plants, including all algae, take up nitrate (NO_3^-) and ammonium (NH_4+) from aquatic environments to meet their needs for water. However, a process known as nitrogen fixation allows some algae, including *Aphanizomenon, Anabaena,* and *Nostoc*, to take in atmospheric nitrogen (N_2), dissolve in water, and produce ammonium. Due to their capacity to absorb and store excess phosphorous, blue-green algae are ideally suited to environments where there is a phosphate shortage.

11.4 Trophic Status Indicator of Aquatic Ecosystem

The trophic state of the lake can be assessed by the abundance and density of algae. The lake's trophic status might serve as a sign of the lake's stage in the eutrophication, or natural ageing, process. The trophic status of the lake can be related to the productivity of the lake and the species makeup of phytoplanktonic algae in the epilimnion.

Oligotrophic: The dominance of *Staurastrum, Closterium,* and *Selenastrum* of some desmids and *Euglena* and some Dinophyta, such as *Ceratium* and *Peridinium*, is a sign that the aquatic ecosystem has low nutrient levels and may even have high oxygen levels.

Mesotrophic: Bacillariophtes, blue-green algae, and green algae predominate in the early and middle summer (Diatoms).

Eutrophic: During the middle of the summer, there may be a high nutrient load, low oxygen levels, and a predominance of cyanophycea like *Anabaena, Aphanizomenon*, and *Microsystis*.

Hypertrophic: Due to high levels of organic manure in artificial growth ponds, the condition is known as hypertrophic. Small, short-lived unicellular algae predominate in this environment.

11.5 Biological Water Purifier

Algae have recently emerged as important organisms for the biological filtering of wastewater, due to their ability to bioaccumulate plant nutrients, heavy metals, pesticides, organic and inorganic hazardous chemicals, and radioactive elements in their cells and bodies. Due to their mutually beneficial relationship, bacteria and algae in the aquatic ecosystem release carbon dioxide and nutrients when organic matter is aerobically oxidised by bacteria. The amount of nutrients in the aquatic environment can be greatly reduced when algae utilize nitrogen and phosphorus for development. Thus, algae play a crucial role in the process of treating wastewater. Through photosynthesis, the amount of dissolvent oxygen increases, and the pH has an impact on the removal of phosphorus, ammonium, and hydrogen sulphide. While physical, chemical, and biological processes as well as mechanical procedures like aeration are used to treat wastewater in ponds, some ponds are entirely dependent on the processes of nature. Algal species differences in heavy metal removal effectiveness are evident. In fact, numerous authors demonstrated the effective removal of lead by *Chlamydomonas*, cadmium, copper, and zinc by *Oscillatoria*, molybdenum by *Scenedesmus chlorelloides*, and zinc by *Chlorella vulgaris.*

11.5.1 Bioindicator

A biological indicator is a specific group of species or a community in the environment at a specific location, at a specified time, with knowledge of physical and chemical elements (bioindicator). The diverse class of organisms known as algae has a vast geographic and temporal range as well as a high level of environmental tolerance. Algae are a crucial component of aquatic ecosystems because of their nutrient requirements, quick reproduction rates, short lifespans, ability to quickly adapt to changes in the community structure of their species, and capacity to increase in density in response to even the slightest change in the water chemistry of their environment. Therefore, algae can be very important in the biological monitoring of water quality. When Kolenti and Cohn (1848) noticed that the biota in polluted rivers was different from that in clean streams, they were the first to speculate that the composition of freshwater algae would change as a result of shifting environmental conditions (Liebmann 1962). Most cyanobacteria thrive in the warmer temperatures and better nutritional conditions of early summer. The colonial blue-green algae *Microsystis* and *Anabaena* regularly show significant summer blooms, which is a sign that the aquatic environment is eutrophic (rich in organic nutrients). Recent advances in water quality monitoring and ecological evaluation have increased interest in phytoplankton abundance, growth, and community composition, as well as photopigments as phytoplankton taxonomic groups and biomass (Paerl et al. 2005).

11.6 Biotechnological Potential of Phytoplankton

Nowadays, There are numerous commercial applications of microalgae in the field of animal nutrition, microalgae in biofertiliser, pharemaceuticals, biofuels, carbon sequestration etc.

11.6.1 Animal Nutrition

Microalgae are essential for the production of high-quality animal feeds for farm animals and aquaculture. Algal biomass is suitable as a beneficial feed addition or replacement for traditional animal feed sources, according to extensive nutritional and toxicological analyses (Dhargalkar and Verlecar 2009). Bioactive chemicals, vital amino acids, pigments, and vitamins are just a few of the valuable substances that microalgae generate. Chlorophylls a, b, and c, E-carotene, Phycocyanin, Xanthophylls (Astaxanthin, Canthaxanthin, and Lutein), and Phycoerythrin are the main pigments. There is an increasing demand for these pigments' use as natural textile colours and printing dyes, in addition to their present applications in food, feed, medicines, and cosmetics. To produce bivalve molluscs, young abalone, crabs, and fin fish species as well as zooplankton for use in food chains, aquaculture uses microalgae as food and live feed. Microalgae are commonly used as a feedstock in the commercial hatchery development of juvenile marine fish and shellfish. Based on their nutritional and nontoxic performance, the marine cyanobacteria *Phorphidium valderianum* were successfully used as feed for aquaculture (Thajuddin and Subramanian 2005). Additionally, dried microalgae biomass is also widely used as an ingredient in formulated feeds for aquaculture species. The potential of *Arthrospira* (Spirulina) as ingredients in animal feed was studied (Belay et al. 1996).

11.6.2 Biofuels

Microalgae have been studied for many years for the production of hydrogen (H_2), methane, oils (triglycerides and hydrocarbons, for biodiesel, jet fuels, etc.), and bioethanol. The main focus of this research was on the generation of biodiesel from microalgae that accumulate large amounts of oil (Sheehan et al. 1998). Because of their high levels of lipids, carbs, and proteins and low levels of lignin, microalgae are regarded as promising substrates for the production of biogas. The following is a list of some selected microalgae species' potential as biofuels (Table 11.2).

11.6.3 Carbon Sequestration

The rising amounts of various greenhouse gases, such as CO_2, emitted through the burning of fossil fuels, have raised awareness of global warming on a global scale. It has been thought of using marine microalgae for biological CO_2 sequestration. One rationale is the possibility of directly using seawater as a growing medium,

Table 11.2 List of some selected microalgae species' potential as biofuels

Name of algae	Content of biofuel percentage	References
Tetraselmis suecica	15–32	Chisti (2007)
Dunaliella sp.	66.6	Shenbaga Devi et al. (2012)
Nannocloropsis sp.	68.5	Shenbaga Devi et al. (2012)
Skeletonema costatum	87	Rekha et al. (2012)
Chlorella marina	70.6	Rekha et al. (2012)
Nannocloropsis sp.		Sheehan et al. (1998)
Nannocloropsis sp.		Hu et al. (2006)

Table 11.3 List of carbon capturing efficacy of some microalgae

Name of algae	CO_2 concentration maximum (%)	References
Scenedesmus sp.	80	Hanagata et al. (1992)
Euglena gracilis	45	Nakano et al. (1996)
Chlorella sp.	40	Hanagata et al. (1992)
Cyanidium caldarium	100	Seckbach et al. (1971)
Chlorococcum littorale	60	Kodama et al. (1993)

which would lower the maintenance expenses of microalgae culture. The coastal region is home to numerous CO_2 sources, including power stations. A number of marine algal species have been tested for CO_2 sequestration applications. Carbon capturing efficacy of some microalgae are listed below (Table 11.3).

11.6.4 Functions of Zooplankton

11.6.4.1 Food Source

In the ecological and food chain of a lake, zooplankton is crucial. They are a heterotrophic part of the planktonic community, acting as a bridge between planktivorous fish and algae or phytoplankton (autotrophs). As zooplanktons are unable to create their own food, they are in charge of eating millions of tiny algae and preventing an algae bloom in the aquatic environment. The majority of zooplankton species are filter feeders and can filter a lake's worth of water in a short amount of time. Many fish and shellfish larvae and fingerlings need live feed right after the start of external feeding because of their broad range of size variation. Numerous studies demonstrate that employing zooplankton as live feed improves both the quality and growth performance of the organisms. The total fish productivity in a lake or other aquatic system can be affected by the amount of zooplankton present. Planktivorous fish and other organisms rely on zooplankton as a key food source. In both freshwater and saltwater environments, the success of some commercial fisheries can be influenced by the presence or absence of robust zooplankton populations. We can safeguard higher ordered organisms, such as fish, whales, and even us humans, by ensuring the health of the lowest rungs of the food chain.

11.7 Bioindicator of Ecological Condition

The word "bioindicator" refers to all sources of abiotic and biotic reactions to ecological changes (Parmar et al. 2016). Bioindicators have a great deal of potential to forecast an ecosystem's natural state or level of contamination (Khatri and Tyagi 2015). According to research on plankton, the planktonic community may be a more accurate predictor of trophic status than changes in chemical and physical characteristics. Cladocerans with widespread adaptations like *Moina micrura* and *Bosmina longirostris*, as well as their relative abundance, are thought of as eutrophic indicators. *Bosmina berolinensis* and copepods are indicators of the trophic

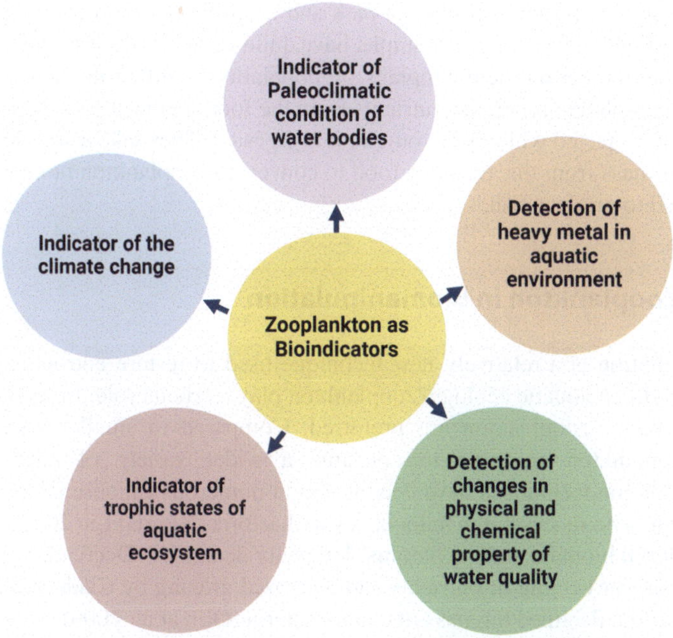

Fig. 11.1 Different ecological indications by zooplankton community

status of the aquatic system in oligo-mesotrophic environments. According to Gannon and Stemberger (1978), the ratio of calanoid copepods to cyclopoid copepods, or N_{Cal}/N_{Cycl}, is a reliable predictor of trophic circumstances. Increased eutrophication typically causes the ratio N_{Cal}/N_{Cycl} to drop (Gannon and Stemberger 1978; Gulati 1983; Caramujo and Boavida 2000). Zooplanktons play a significant role in trophic interactions, such as the trophic food chain and food webs, energy transmission, and overall ecological stability, making them an essential part of aquatic ecosystems. In addition to these crucial ecological characteristics, they also play a dynamic role in pollutant transfer, bioaccumulation, and biomonitoring phenomena. As a result, polluted zones in the aquatic part of the world are correctly identified and analysed at an early stage, allowing for the proper application of corrective measures aimed at their restoration. Since zooplankton assemblages and their ecology depend on both biotic and abiotic factors, their entire function as bioindicators can be completely established by taking into account all of their influencing elements (Fig. 11.1).

11.8 Control of Eutrophication

To reduce eutrophication, Cladoceran (*Bosmina longirostris* and *Daphnia lumholtzi*) limits the algal population. Zooplankton is used to track the aquatic ecosystem's recovery phase. Numerous studies have shown how differently Cladocerans, cyclopoid copepods, and rotifers can pick and catch food within the zooplankton. Cladocerans and copepods, for example, have quite varied feeding strategies, which may have an impact on the phytoplankton population in different ways. No living thing can completely retain the nutrients from the food it consumes. The remainder of the meal is ejected while only some of it is digested. They take a sizable quantity of algal biomass from the available food to convert to zooplankton biomass, which is then used to produce fish.

11.9 Zooplankton in Biomanipulation

Biomanipulation is a relatively new technique used to restore eutrophied aquatic ecosystem. In an aquatic ecology, zooplankton play a crucial role. In terms of algal filtration, larger zooplankton are preferred. Compared to smaller zooplankton, larger zooplankton are able to consume a wider variety of algal species. Biomanipulation techniques have been tested in numerous Chinese lakes and have been shown to be successful in reducing cyanobacteria (Xie and Liu 2001). Both the cyanobacterial biomass and the blooms' longevity decreased. Decreased concentrations and slower cycling of nutrients and increased grazing by Cladocerans probably affected the declined biomass of cyanobacteria (Olin et al. 2006).

11.10 Biotechnological Potential of Zooplankton

1. Live feed.
2. Green fluorescent protein (GFP).
3. Astaxanthin.

11.10.1 Live Feed

Water fleas such as Daphnia are microscopic planktonic crustaceans. The animals belong to the Cladocera group. These shelled crustaceans range in size from 0.2 to 5.0 mm. The *Daphnia pulex* and *Daphnia magna* species are the most prevalent ones. The most widespread of all these species is *Daphnia pulex*. The largest species of Daphnia, *Daphnia magna*, can grow to a maximum size of 5 mm. Both species are commonly used as fish food. Water fleas are a common live food, just like other zooplankton. This is a good option for larger fish fry or preparing adult fish for spawning. Cyclops are microscopic planktonic crustaceans, belonging to the Copepoda group. The size of the body of the Copepods is 0.3–3.5 mm. Copepods that have been cultured to grow fish larvae from a variety of marine species. Calanoida, harpacticoida, and cyclopoida are the three main orders whose cultures are emphasised by the live feed business. The seven other orders cannot be used to meet life needs since they are parasitic. Numerous marine fish larvae require food that is between 50 and 100 pm broad when they first begin to feed. Copepods are

Table 11.4 Biotic indices for assessment of trophic state of aquatic ecosystem

Rotifer Trophic State Index	References
$TSI_{ROT1} = 5.38\ln(Nr) + 19.28$; where Nr number of rotifer (ind/l)	Ejsmont-Karabin (2012)
$TSI_{ROT2} = 5.38\ln(B) + 64.47$; where B biomass of rotifer (mg ww/l)	
$TSI_{ROT3} = 3.85(B/Nr)^{-0.318}$	
$TSI_{ROT4} = 0.144$ TECTA +54.8; percentage of spineless form (tecta) in abundance of *Keratella cochlearis*	
$TSI_{ROT5} = 0.203$IHT+ 40.0; number of species contributed to the high trophic status indicator group	
Crustacean trophic state index	
$TSI_{CR1} = 25.5\ Nr^{0.142}$, where Nr number of crustacean abundance (ind/l)	Ejsmont-Karabin and Karabin (2013)
$TSI_{CR2} = 57.6B^{0.081}$; Where B total wet cyclopoid biomass (mg/l)	
$TSI_{CR3} = 40.9CB^{0.097}$; Percentage of cyclopoid biomass in the total crustacean biomass	
$TSI_{CR4} = 58.3(CY/CL)^{0.071}$; CY/CL: Ratio of cyclopoids (CY) to cladoceran biomass (CL)	
$TSI_{CR5} = 5.08\mathrm{Ln}(CY/CA) + 46.6$; ratio of cyclopoids (CY) to calanoid biomass (CA)	

available in a wide range of sizes, species, and characteristics. They also include significant amounts of protein, highly unsaturated fatty acids (HUFA), carotenoids, and other vital substances. Copepods are increasingly being used in the live feed business due to their greater concentrations of DHA and EPA.

11.10.2 Assess Trophic State of Aquatic Ecosystem

One of the biotic elements used to assess the health of aquatic ecosystems is rotifer (Ejsmont-Karabin (2012). For the rotifer trophic status index, the rotifer community structure has been used (TSI_{ROT}). Ejsmont-Karabin and Karabin (2013) developed Crustacean-based indices (TSI_{CR}) for the assessment of Crustacean-based trophic status indices (Table 11.4).

References

Belay A, Kato T, Ota Y (1996) Spirulina (Arthrospira): potential application as an animal feed supplement. J Appl Phycol 8:303–311

Caramujo MJ, Boavida MJ (2000) The crustacean communities of river Tagus Reservoirs. Zooplankton structure as reservoir trophic state indicator. Limnetica 18:37–56

Chisti Y (2007) Biodiesel from microalgae beats bioethanol. Trends Biotechnol 25:126–131

Datta-Munshi J, Roy SP, Datta-Munshi JS (2010) Manual of fresh—water biota. Narendra Publishing House, Delhi, p 435

Dhargalkar VK, Verlecar XN (2009) Southern Ocean seaweeds: a resource for exploration in food and drugs. Aquaculture 257:229–242

Ejsmont-Karabin J (2012) The usefulness of zooplankton as lake ecosystem indicators: rotifer trophic index. Pol J Ecol 60:339–350

Ejsmont-Karabin J, Karabin A (2013) The suitability of zooplankton as lake ecosystem indicators: crustacean trophic state index. Polish J Ecology 61:561–573

Gannon JE, Stemberger RS (1978) Zooplankton (especially crustacean and rotifers) as indicators of water quality. T. Am. Micros. Soc. 97:16–35

Gulati RD (1983) Zooplankton and its grazing as indicators of trophic status in Dutch lakes. Environmental Monitoring and Assessment 3:343–354

Hanagata N, Takeuchi T, Fukuju Y, Barnes DJ, Karube I (1992) Tolerance of microalgae to high CO2 and high temperature. Phytochem 31(10):3345–3348

Hoffmann L (1989) Algae of terrestrial habitats. Bot Rev 55:77–105

Hu Q, Zhang CW, Sommerfeld M (2006) Biodiesel from Algae: Lessons Learned Over the Past 60 Years and Future Perspectives. Juneau, Alaska: Annual Meeting of the Phycological Society of America, July 7–12, pp. 40–41 (Abstract)

Khatri N, Tyagi S (2015) Influences of natural and anthropogenic factors on surface and groundwater quality in rural and urban areas. Front Life Sci 8:23–39

Kodama M, Ikemoto, Miyachi S (1993) A new species of highly CO2-tolerant fast-growing marine microalga suitable for high-density culture, J. Mar. Biotechnol. 1: 21–25

Liebmann H (1962) Handbunch der frishwasser und Abwasser-biologie. R. Oldenbourg, Munchen, p 113

Nakano Y, Miyatake K, Okuno H, Hamazaki K et al (1996) Growth of photosynthetic algae Euglena in high CO2 conditions and its photosynthetic characteristics. Acta Horticulturae 440(9):49–54

Paerl HW, Dyble J, Pinchney JL et al (2005) Using microalgal indicators to assess humane and climate-induced ecological change in estuaries. In: Bortone SA (ed) Estuarine indicators. CRC Press, Boca Raton, FL, pp 145–174

Parmar TK, Rawtani D, Agrawal YK (2016) Bioindicators: the natural indicator of environmental pollution. Front Life Sci 9:110–118

Rekha V, Gurusamy R, Santhanam P, Shenbaga DA, Ananth S (2012) Culture and biofuel production efficiency of marine microalgae Chlorella marina and Skeletonema costatum. Indian J Mar Sci 41:152–158

Round FE, Crawford RM, Mann DG (1990) The diatoms: biology and morphology of the genera. Cambridge University Press, Cambridge, p 747

Seckbach J, Gross H, Nathan MB (1971) Growth and Photosynthesis of Cyanidium caldarium cultured under pure CO2. Israel J Bot 20:84–90

Sheehan J, Dunahay T (1998) A Look Back at the U.S. Department of Energy's Aquatic Species Program-Biodiesel from Algae, National Renewable Energy Program

Shenbaga D, Santhanam A, Rekha P, Ananth V, Prasath S, Nandakumar BB, Jeyanthi RS, Kumar SD (2012) Culture and biofuel producing efficacy of marine microalgae Dunaliella salina and Nannochloropsis sp. J Algal Biomass Utln 3(4):38–44

Thajuddin N, Subramanian G (2005) Cyanobacterial biodiversity and potential applications in biotechnology. Curr Sci 89:47–57

Xie P, Jiankang L (2001) Practical success of biomanipulation using filter-feeding fish to control cyanobacteria blooms: a synthesis of decades of research and application in a subtropical hypereutrophic lake. Sci World J 1:337–356

Xu F-L, Dawson RW, Tao S, Cao J, Li B-G (2001) A method for lake ecosystem health assessment: an ecological modeling method (EMM) and its application. Hydrobiologia 443:159–175

Surrogacy Technology in Fisheries and Aquaculture

12

Bijay Kumar Behera

Abstract

The surrogacy technology in fisheries and aquaculture emphasizes in improving breeding efficiency, genetic diversity, and sustainability. Surrogacy technology replaces natural reproductive processes in aquatic species with advanced biotechnological methods like genetic engineering, embryonic manipulation, hormonal control, and broodstock management. This shift to sophisticated genetic interventions, including CRISPR-Cas9, allows for precise genetic enhancements. Genetic engineering is central to introducing traits like disease resistance and accelerated growth. Embryonic manipulation techniques, such as microinjection and somatic cell nuclear transfer, along with hormonal control for spawning induction and selective breeding in broodstock management, are essential tools in modern aquaculture and genetic research. These techniques enable precise genetic modifications, enhance reproductive efficiency, and improve desirable traits in cultured species, thereby contributing to sustainable and productive aquaculture practices. Different types of surrogacy, including interspecies embryonic transfer and germ-line surrogacy, are analysed for their role in genetic diversity and species recovery. Case studies also illustrate their applications in enhancing productivity, conservation, and understanding of aquatic biology. Future perspectives, involving emerging trends like artificial intelligence and advanced gene editing, are discussed, highlighting their potential in revolutionizing fisheries and aquaculture for ecological responsibility and sustainability. This overview underscores the significance of surrogacy technology in global food security and conservation efforts.

B. K. Behera (✉)
College of Fisheries, Rani Lakshmi Bai Central Agricultural University, Jhansi, Uttar Pradesh, India

B. K. Behera (ed.), *Current Trends in Fisheries Biotechnology*,
https://doi.org/10.1007/978-981-97-3165-7_12

Keywords

Aquaculture · Conservation · CRISPR · Fisheries · Surrogacy

12.1 Introduction

Surrogacy technology in fisheries and aquaculture refers to advanced biotechnological methods employed to substitute natural reproductive processes in aquatic species. This technology encompasses various techniques designed to artificially replicate or enhance the breeding and development of fish and other aquatic organisms. Typically, it involves the manipulation of genetic material, embryonic development, and hormonal control to ensure successful reproduction and growth of species under controlled conditions. The primary objective of surrogacy technology is to augment the efficiency and yield of aquaculture operations, conserve endangered species, and support sustainable fisheries management (Mobrand et al. 2005; Mitra et al. 2023). This scientific approach integrates principles from genetics, molecular biology, and reproductive physiology, aiming to optimize aquatic populations for commercial, conservation, or research purposes.

The historical evolution of surrogacy technology in fisheries and aquaculture is a testament to the progressive integration of advanced biological techniques in the management and cultivation of aquatic species. Initially, traditional breeding methods dominated, focusing primarily on selective breeding and environmental manipulation to enhance yield and resilience (Donaldson and Olson 1957; Donaldson 1971; Garber and Sullivan 2006; Gjedrem et al. 2012; Chavanne et al. 2016). As scientific understanding deepened, the late twentieth century witnessed the advent of more sophisticated methodologies. This era marked the commencement of genetic intervention techniques, heralding a new phase in aquaculture and fisheries management. Genetic engineering and biotechnology began to play pivotal roles, leading to the development of transgenic fish and the introduction of specific traits to improve growth rates, disease resistance, and environmental adaptability (Gui et al. 2022). The turn of the millennium saw a significant leap with the incorporation of molecular biology and genomics into surrogacy technology. These advancements facilitated precise genetic modifications and the development of gene editing tools like CRISPR-Cas9, revolutionizing the ability to engineer aquatic species at the molecular level.

Concurrently, there was a growing emphasis on sustainability and conservation. Surrogacy technology adapted to these needs by focusing on the restoration of endangered species and the establishment of more sustainable aquaculture practices. Techniques such as surrogate broodstock technology, where one species carries the embryos of another, emerged, offering innovative solutions for species conservation. Today, surrogacy technology in fisheries and aquaculture stands at the forefront of biotechnological innovation, blending traditional practices with cutting-edge genetic and reproductive technologies. This evolution reflects a continuous effort to harmonize economic productivity with ecological responsibility in the aquatic sector.

The significance and pertinence of surrogacy technology in aquaculture and fisheries practices are underscored by its multifaceted contributions to the field. This technology plays a crucial role in addressing the burgeoning demand for aquatic resources while fostering sustainable and efficient practices. Primarily, surrogacy technology enhances reproductive efficiency and genetic diversity in aquaculture (Jin et al. 2021). By facilitating controlled breeding and genetic manipulation, it enables the production of robust and disease-resistant species, which are essential for maintaining high-yield and sustainable aquaculture systems. This aspect is particularly crucial in the context of global food security, as aquaculture represents a rapidly growing source of protein for human consumption. In the realm of conservation, surrogacy technology offers a vital tool for the preservation of endangered species (Majhi et al. 2009; Cowl et al. 2023). By enabling the reproduction of threatened species in controlled environments, it aids in replenishing dwindling populations and restoring ecological balance. Moreover, the technology allows for the study and conservation of genetic material, ensuring the long-term preservation of biodiversity in aquatic ecosystems.

Additionally, surrogacy technology contributes to the field of fisheries management by providing methods for stock enhancement and habitat restoration (Jin et al. 2021). By producing species that can be reintroduced into their natural habitats, it assists in mitigating the impacts of overfishing and environmental degradation. Furthermore, the application of surrogacy technology in research opens new avenues for understanding aquatic biology and genetics. It facilitates the study of developmental processes, genetic traits, and the impacts of environmental changes on aquatic species, thereby enriching scientific knowledge and informing better management practices.

12.2 Basic Principles of Surrogacy Technology

12.2.1 Explanation of the Underlying Biological and Technological Principles

Surrogacy technology in fisheries and aquaculture is underpinned by a confluence of biological and technological principles that facilitate the artificial reproduction and cultivation of aquatic species. These have also been summarized in Fig. 12.1.

12.2.1.1 Genetic Engineering

At its core, surrogacy technology often involves genetic manipulation to introduce or modify specific traits in aquatic organisms. This is achieved through techniques like gene editing, where tools like CRISPR-Cas9 are used to precisely alter the DNA sequence, enabling the enhancement of desirable traits such as growth rate, disease resistance, or environmental adaptability.

Genetic engineering within surrogacy technology in fisheries and aquaculture encompasses a series of complex biological and technological processes aimed at manipulating the genetic makeup of aquatic organisms (Jin et al. 2021; Franěk et al. 2022). This technology hinges on the precise alteration of an organism's DNA through advanced techniques such as CRISPR-Cas9, enabling targeted genetic modifications. These modifications can introduce or suppress specific traits, such as improved growth rates or enhanced disease resistance, crucial for the development

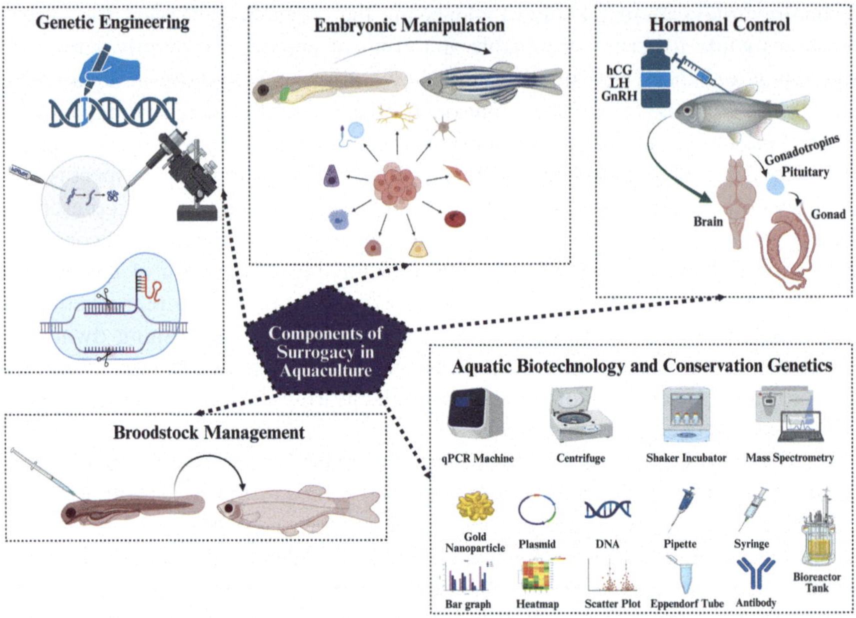

Fig. 12.1 Biological principles and different components of surrogacy technology

of robust aquaculture species. Transgenesis, another key facet of genetic engineering in surrogacy, involves the transfer of genes across species. This process allows a surrogate species to express desirable characteristics in its offspring, effectively enhancing certain phenotypic traits. Molecular cloning complements this by facilitating the replication and analysis of specific genes, thereby providing deeper insight into genetic influences on traits.

Synthetic biology extends these capabilities further, enabling the design and construction of novel biological systems. Through this, organisms can be engineered with new properties, expanding the potential applications of surrogacy in aquaculture. The practice of genomic selection plays a pivotal role in refining breeding programs. By selecting individuals based on genetic markers, it becomes possible to enhance traits that are otherwise challenging to measure or that manifest later in life. Finally, germ-line modification is a crucial component in ensuring the heritability of engineered traits. By altering the genetic information in an organism's germ cells, desired changes are passed down to future generations, solidifying the role of genetic engineering in producing optimally adapted aquatic species.

This is how the genetic engineering principles in surrogacy technology represent a confluence of cutting-edge genetic manipulation techniques. These principles enable the strategic enhancement of aquaculture species, leading to improved productivity and sustainability in aquaculture and fisheries management.

12.2.1.2 Embryonic Manipulation

Embryonic manipulation in surrogacy within the realms of fisheries and aquaculture is a process that integrates biological and technological insights to control and guide the early development stages of aquatic organisms. This technique fundamentally revolves around the strategic intervention in the embryonic stage of an organism, aiming to enhance or alter developmental outcomes in a controlled manner.

At the heart of embryonic manipulation is the transfer of embryos from a donor species to a surrogate species. This process, known as interspecies embryonic transfer, allows a surrogate organism to carry and nurture the embryos of another species, often used in cases where the donor species faces challenges in natural reproduction or is at risk of extinction (Yoshizaki and Lee 2018). The surrogate thus acts as a biological incubator, providing the necessary environmental conditions for the embryo to develop and eventually hatch. An outline of this method through cryopreservation in target of species survival and conservation has been illustrated in Fig. 12.2. The process also involves in vitro fertilization (IVF), where eggs are fertilized outside the organism in a controlled laboratory setting. IVF is crucial in ensuring the successful union of gametes, particularly in species where natural fertilization is inefficient or impossible under aquaculture conditions.

Furthermore, embryonic manipulation includes techniques such as microinjection and somatic cell nuclear transfer (SCNT). Microinjection involves the insertion of substances, such as DNA or RNA, directly into the embryo, allowing for genetic modifications at a very early stage. SCNT, on the other hand, is a method of cloning

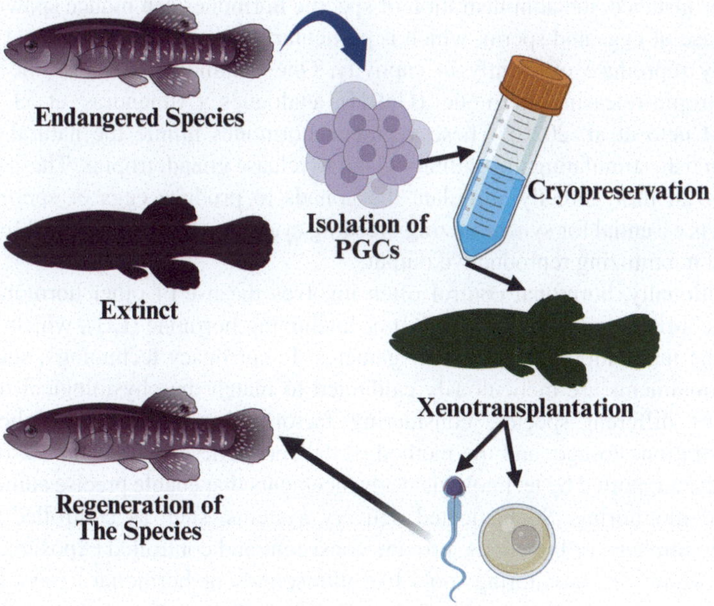

Fig. 12.2 The process of surrogacy technology in conservation of species

where the nucleus of a somatic cell is transferred to an egg cell, facilitating the creation of genetically identical organisms.

These practices in embryonic manipulation are supported by a range of technological advancements, including microscopy for detailed observation and manipulation, incubation systems that mimic natural embryonic environments, and precise instruments for handling delicate embryonic materials. In essence, embryonic manipulation in surrogacy technology leverages a deep understanding of developmental biology and advanced technological tools to influence and guide the embryonic development of aquatic species. This approach is pivotal in enhancing breeding programs, conserving endangered species, and improving the overall efficiency and sustainability of aquaculture practices.

12.2.1.3 Hormonal Control

Hormonal treatments are used to induce spawning or the release of eggs and sperm in fish, which is crucial for controlled breeding. This technique is essential in synchronizing breeding cycles and ensuring successful fertilization, especially in species that are difficult to breed in captivity (Mañanós et al. 2008; Mylonas et al. 2010). Hormonal control in surrogacy technology, particularly within fisheries and aquaculture, involves the strategic use of hormones to regulate and stimulate the reproductive processes of aquatic organisms. This facet of surrogacy technology is grounded in endocrinology, the branch of biology concerned with the study of hormones and their effects on organisms.

The fundamental principle here is to manipulate the hormonal cues that trigger reproductive behaviours and physiological changes in fish and other aquatic species. For instance, the administration of specific hormones can induce spawning or the release of eggs and sperm, which is particularly valuable in species that do not naturally reproduce efficiently in captivity. One common practice is the use of gonadotropin-releasing hormone (GnRH) analogues (Almendras et al. 1988; Muñoz-Cueto et al. 2020). These synthetic hormones mimic the natural GnRH found in fish, stimulating the pituitary gland to release gonadotropins. These gonadotropins, in turn, directly stimulate the gonads to produce eggs or sperm. This method is essential for synchronizing breeding cycles, ensuring successful fertilization and maximizing reproductive output.

Additionally, hormonal control often involves the use of other hormones like human chorionic gonadotropin (hCG) or luteinizing hormone (LH), which further aid in the maturation and release of gametes. In surrogacy technology, such hormonal treatments are meticulously calibrated to match the physiological requirements of different species, considering factors like the timing of hormone administration, dosage, and the method of delivery. These hormonal control techniques are supported by technological advancements that enable precise administration and monitoring. Sophisticated delivery systems, such as controlled-release hormone implants or injections, provide consistent and controlled exposure to hormones. Moreover, monitoring tools like ultrasounds or hormonal assays help in assessing the reproductive status and response of aquatic organisms to the treatments.

So hormonal control in surrogacy technology is a critical aspect that leverages endocrinological principles and advanced technologies. This approach ensures effective reproduction management in aquaculture, enhancing breeding success and supporting the sustainable production of aquatic species.

12.2.1.4 Broodstock Management

Broodstock management in surrogacy technology represents a vital component of reproductive biotechnology, focusing on the selection, maintenance, and breeding of mature fish or other aquatic species for the purpose of producing offspring. This domain combines biological principles with technological advancements to optimize the health, genetic diversity, and reproductive efficiency of broodstock populations.

At the biological level, broodstock management involves selecting individuals with desirable genetic traits. This selection is based on criteria like growth rate, disease resistance, and other phenotypic characteristics relevant to aquaculture productivity and sustainability. The aim is to ensure the propagation of these advantageous as well as commercially important traits in future generations, thereby enhancing the overall quality of the stock. The process of germ cell transplantation in fisheries and aquaculture offers a compelling approach to amplify the presence of superior traits within fish populations (Yoshizaki and Yazawa 2019). For example, in Fig. 12.3, the fish possessing superior traits, represented by black colour, serve as donors, while the recipients, depicted in green, become the carriers of these enhanced genetic qualities. The essence of this technique lies in its ability to use a single donor fish, endowed with superior traits, to generate a substantial number of

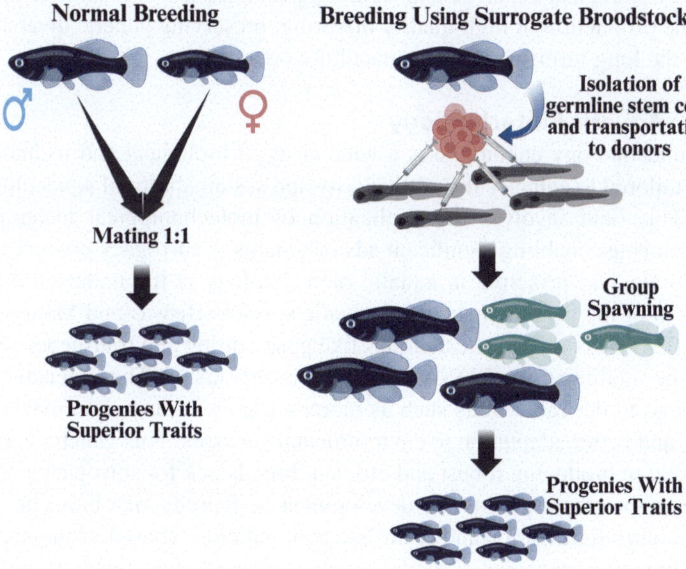

Fig. 12.3 Outline of the method of broodstock technology in fisheries and aquaculture

recipient fish. This multiplication effect significantly elevates the likelihood that the gametes produced by these recipients will carry the genetic makeup originating from the superior donor. As these recipient fish engage in group spawning, any mating between a female and a male recipient can lead to fertilization events where both the eggs and sperm are derived from donors with superior traits. This method effectively disseminates the desirable characteristics throughout the population, as the offspring resulting from such mating will inherit the enhanced genetic qualities. Thus, germ cell transplantation stands as a powerful tool in aquaculture, allowing for the strategic propagation of beneficial traits, which in turn can lead to improvements in growth, resilience, and overall productivity of the fish populations.

From a health perspective, maintaining the optimal physical condition of broodstock is crucial. This involves providing a suitable environment that mimics natural habitats as closely as possible, including regulated water quality, temperature, and nutrition. Ensuring the well-being of broodstock is essential for their reproductive success and the health of their progeny. Technologically, broodstock management incorporates various tools and methods. Sophisticated monitoring systems are employed to track water parameters and fish health, while genetic analysis techniques, such as DNA sequencing and genotyping, are used to understand the genetic makeup of the broodstock. These technologies enable precise decision-making in the selection and breeding processes. Moreover, reproductive technologies, including artificial insemination and hormone-induced spawning, are frequently utilized in broodstock management. These methods allow for controlled breeding, overcoming limitations that might occur in natural reproductive behaviours, especially in a captive environment.

In essence, broodstock management in surrogacy technology represents an intricate blend of biological expertise and technological innovation. It is fundamental in ensuring the production of high-quality offspring, preserving genetic diversity, and sustaining the long-term success of aquaculture operations.

12.2.1.5 Aquatic Biotechnology

Aquatic biotechnology encompasses a wide array of biological and technological principles tailored to enhance the productivity and sustainability of aquaculture and fisheries. This field involves the application of biotechnological techniques to aquatic organisms, enabling significant advancements in surrogacy practices.

A key biological principle in aquatic biotechnology is the understanding and manipulation of the genetic makeup of aquatic species (Biswas and Maurye 2017; Dunham 2023). This involves techniques like gene editing and transgenesis, which allow for the modification of DNA sequences in organisms. These genetic alterations can lead to desirable traits such as increased growth rates, improved disease resistance, and better adaptation to environmental stressors. This genetic manipulation is crucial in producing robust and efficient broodstock for surrogacy purposes.

Another important aspect is the development of reproductive biotechnologies, including in vitro fertilization and germ-line manipulation. These techniques enable controlled breeding processes, essential for maintaining genetic diversity and health in captive aquatic populations. By manipulating the reproductive processes, surrogacy technology can overcome natural breeding limitations in certain species, thereby enhancing breeding success. From a technological standpoint, aquatic biotechnology employs advanced tools for monitoring and controlling the aquatic environment. This includes precision aquaculture technologies, like automated feeding systems, water quality sensors, and real-time monitoring of aquatic conditions. These technologies ensure optimal growth conditions, crucial for the health and development of surrogate broodstock and their offspring.

Additionally, bioreactors and other cultivation systems are used for producing microalgae and other feed components. These systems provide a sustainable and efficient means of producing high-quality feed, integral to the nutrition of aquatic organisms in surrogacy programs. Aquatic biotechnology also integrates bioinformatics and computational tools for the analysis of genetic data and the modelling of biological processes. These tools are vital for understanding the complex interactions within aquatic ecosystems and for making informed decisions in the management of surrogacy programs.

This is how the aquatic biotechnology in surrogacy technology combines advanced genetic manipulation, reproductive techniques, and environmental control technologies. These elements work synergistically to improve the efficiency, sustainability, and productivity of aquaculture and fisheries, playing a crucial role in the advancement of surrogacy practices in these sectors.

12.2.1.6 Conservation Genetics

Conservation genetics within surrogacy technology in fisheries and aquaculture is a discipline that merges genetic principles with conservation efforts, focusing on preserving the genetic diversity and viability of aquatic species. This field plays a critical role in surrogacy by ensuring that the genetic makeup of both wild and cultivated populations is maintained, aiding in the sustainability and ecological balance of aquatic environments.

Central to conservation genetics is the understanding of genetic variation within and between populations of species. This involves analysing genetic markers, which are specific DNA sequences used to identify genetic differences and similarities. By assessing these markers, scientists can gauge the level of genetic diversity in a population, which is crucial for its health and ability to adapt to changing environments (Leberg 1992; McCusker and Bentzen 2010). High genetic diversity is generally associated with greater resilience to diseases, environmental changes, and other stressors.

Another key aspect is the management of genetic resources, which includes the preservation of unique genetic traits and the prevention of inbreeding. In surrogacy programs, careful selection and breeding strategies are employed to maintain a healthy level of genetic variation, avoiding the genetic homogenization that can occur in captive breeding. Technological advancements in molecular biology and genomics significantly bolster conservation genetics efforts. Tools like DNA sequencing and genomic analysis enable detailed examinations of genetic structures and functions. These technologies provide insights into the evolutionary history and adaptive capabilities of species, guiding effective conservation strategies.

Cryopreservation of gametes and embryos is another crucial technology in conservation genetics. This involves storing genetic material at very low temperatures for future use, a method especially valuable for preserving the genetics of endangered or rare species. By maintaining a repository of genetic material, surrogacy programs can reintroduce or bolster genetic variation in populations as needed. Additionally, bioinformatics and computational modelling are increasingly important in conservation genetics. These tools allow for the simulation of genetic changes over time, predicting the outcomes of different conservation strategies. This predictive capacity is vital for planning and implementing effective surrogacy and conservation programs.

So conservation genetics in surrogacy technology blends deep genetic insights with advanced biotechnological tools to preserve and manage the genetic diversity of aquatic species. This approach is fundamental in ensuring the long-term sustainability and ecological integrity of both natural and managed aquatic ecosystems.

12.2.2 Types of Surrogacy Technology Used in Fisheries and Aquaculture

Surrogacy technology in fisheries and aquaculture encompasses various types, each employing distinct biological and technological methodologies to enhance the production and conservation of aquatic species.

Interspecies Embryonic Transfer is a prominent type where embryos from one species are implanted into the surrogate mother of another species. This technique is particularly useful for conservation efforts, allowing endangered or difficult-to-breed species to reproduce successfully using more common or hardier surrogate species.

Germ-Line Surrogacy involves the transplantation of germ cells (sperm or eggs) from a donor species into a surrogate, where these cells integrate into the surrogate's reproductive system. The surrogate then produces offspring that are genetically related to the donor species. This method is significant for maintaining genetic diversity and assisting in the recovery of species with declining populations.

Artificial Reproductive Technologies (ART), including in vitro fertilization and artificial insemination, are widely used in aquaculture (Gambini et al. 2023). These techniques control the fertilization process, ensuring higher success rates and allowing for genetic selection and manipulation to produce desired traits in offspring. Genetic Engineering and Gene Editing techniques, such as CRISPR-Cas9, enable precise modifications to the DNA of aquatic organisms. By altering specific genes, it is possible to enhance characteristics like growth rate, disease resistance, and environmental adaptability, thereby improving the overall quality and sustainability of aquaculture operations. Broodstock Genetic Management involves selective breeding practices where individuals with desirable traits are chosen as parents for the next generation. This selective process ensures the propagation of advantageous genetic qualities, improving the resilience and productivity of aquaculture species. Hormonal Induction Techniques are used to stimulate and synchronize spawning in fish. By manipulating the hormonal cycles of the broodstock, these techniques facilitate controlled breeding, allowing for efficient and predictable production schedules in aquaculture facilities.

Each of these types of surrogacy technology plays a vital role in modern fisheries and aquaculture. They enable enhanced control over reproduction and genetics, contributing to more sustainable and productive aquatic farming practices, as well as aiding in the conservation of endangered species.

12.3 Applications in Fisheries and Aquaculture

12.3.1 Case Studies of Surrogacy Technology in Fisheries

Several notable case studies illustrate the application of surrogacy technology in fisheries:

Rainbow Trout and Tiger Trout: Researchers at the Hungarian University of Agriculture and Life Sciences conducted a study where rainbow trout germ cells were transplanted into tiger trout larvae (Lujić et al. 2018). This experiment is an example of interspecies germ cell transplantation, a method that holds promise for enhancing genetic diversity and conservation efforts in fisheries.

Cryopreservation and Zebrafish Embryos: In a groundbreaking study in 2017, researchers successfully cryopreserved zebrafish embryos (Khosla et al. 2017). They microinjected the embryos with gold nanoparticles and submerged them in liquid nitrogen, followed by thawing with a laser. This method prevented the formation of damaging ice crystals, a significant challenge in cryopreservation technologies.

Germ Cell Transplantation for Breeding: A method developed by scientist Goro Yoshizaki involves collecting early germ cells from the gonads of a donor fish and transplanting them into another fish, where they develop into sperm or eggs (Yoshizaki and Lee 2018). This approach has been used to breed new fish that genetically match the donor species, and it can be combined with cryopreservation for enhanced efficiency.

Sturgeon Conservation in the Czech Republic: Scientists in the Czech Republic are using surrogacy technology to conserve sturgeon, a critically endangered group of species (Baloch et al. 2019). Early germ cells from the endangered Russian sturgeon were transplanted into sterlet, a more common species. This method aims to accelerate breeding and recovery of sturgeon populations, which are challenging due to their long time to reach sexual maturity.

These case studies demonstrate the diverse applications of surrogacy technology in fisheries, ranging from species conservation to innovative breeding techniques.

12.3.2 Benefits and Advancements Brought by these Technologies

Surrogacy technology in fisheries and aquaculture has brought about a multitude of benefits and advancements, significantly transforming the sector. One of the primary advantages is the enhanced reproductive efficiency it offers. Techniques like artificial insemination and in vitro fertilization have enabled controlled breeding, leading to increased yields and the ability to selectively enhance desirable traits in aquatic species. This has been pivotal in meeting the growing global demand for fish and seafood.

The application of genetic engineering and gene editing, particularly using tools like CRISPR-Cas9, has enabled the precise modification of aquatic species' genomes (Dehler et al. 2016; Liu et al. 2018; Segev-Hadar et al. 2021). This has facilitated the development of species with improved growth rates, better disease resistance, and enhanced adaptability to environmental changes. Such genetic improvements are critical for the sustainability and resilience of aquaculture operations in the face of climate change and disease outbreaks. Conservation of endangered species is another significant achievement of surrogacy technology. Through techniques like germ-line surrogacy and cryopreservation, the genetic diversity of rare or endangered species can be preserved, aiding in their recovery and conservation. This is crucial for maintaining biodiversity and ecological balance in aquatic ecosystems.

Moreover, these technologies have led to a better understanding of aquatic biology and genetics. The insights gained from genetic manipulation and reproductive technologies contribute to more informed management and conservation strategies, enhancing the overall health and sustainability of aquatic environments. Furthermore, surrogacy technology has enabled aquaculture to become more efficient and environmentally sustainable. By improving breeding practices and disease resistance, the reliance on wild catch can be reduced, alleviating the pressure on natural fish populations and contributing to more sustainable fishing practices.

The advancements brought by surrogacy technology in fisheries and aquaculture have been far-reaching, encompassing increased productivity, conservation of biodiversity, enhanced understanding of aquatic biology, and a move towards more sustainable and resilient aquaculture practices. These developments are crucial in addressing global food security, environmental sustainability, and the conservation of aquatic ecosystems.

12.4 Methodological Approaches

12.4.1 Detailed Description of Common Methodologies in Surrogacy Technology

12.4.1.1 Germ Cell Transplantation

This method involves transferring germ cells (sperm or egg precursors) from a donor species to a recipient, which then acts as a surrogate (Lacerda et al. 2013). The surrogate's reproductive system integrates these cells, allowing it to produce offspring genetically identical to the donor species. This methodology is pivotal in conservation efforts, especially for species that are endangered or have difficulties in natural reproduction. The method involves following steps:

1. Donor and Recipient Selection: The initial step involves choosing suitable donor and recipient species. Donors are typically selected based on specific traits or genetic makeup that are desired in the offspring. Recipients are chosen for their compatibility with the donor species and their ability to successfully carry and produce offspring.
2. Collection of Germ Cells: Germ cells (sperm or egg precursors) are harvested from the donor species. This is usually done by dissecting the gonads of the donor to extract spermatogonia or oogonia. The process requires precision to ensure the viability of the collected cells.
3. Preparation of Recipient: The recipient, or surrogate, undergoes preparation to receive the donor germ cells. This often involves conditioning treatments to suppress its own germ cell development, creating a biological niche for the donor cells to integrate successfully.
4. Transplantation of Germ Cells: The harvested germ cells are then transplanted into the recipient. This is usually done via microinjection, where cells are carefully injected into the gonadal tissue of the recipient. The procedure requires meticulous handling and expert use of microsurgical tools to ensure successful transplantation.
5. Integration and Maturation: Post-transplantation, the germ cells integrate into the recipient's reproductive system. Over time, these cells mature and become fully functional gametes—sperm or eggs—of the donor species. This maturation process is closely monitored to ensure proper development.
6. Breeding and Production of Offspring: Once the germ cells have matured, the surrogate is bred, either naturally or through artificial reproductive techniques. The resulting offspring carry the genetic material of the donor species, thus achieving the goals of the transplantation—be it conservation, genetic diversity, or enhanced aquaculture production.
7. Assessment and Monitoring: Post-breeding, the offspring and the surrogate are closely monitored and assessed. Genetic testing is often conducted to confirm the genetic lineage of the offspring. This monitoring ensures the overall success of the transplantation process and the health of the produced offspring.

Each of these steps involves a high degree of biological expertise and technological precision, reflecting the complexity and sophistication of germ cell transplantation in aquatic species management.

12.4.1.2 In Vitro Fertilization (IVF)

IVF is a process where eggs are fertilized by sperm outside the body, typically in a laboratory environment. This technique is essential in controlled breeding programs, allowing for the selection of specific genetic traits and improving the genetic diversity of cultivated species (Gazsi et al. 2021).

It involves several stages, each critical for successfully fertilizing eggs outside the natural aquatic environment and is used for enhancing breeding programs, genetic research, and conservation efforts.

1. Broodstock Selection: The first step involves selecting healthy and genetically desirable male and female broodstock. This selection is based on specific criteria such as genetic quality, health status, and traits like growth rate or disease resistance, crucial for the goals of the IVF process.
2. Gamete Collection: The next stage is the collection of gametes—eggs from females and sperm from males. For females, this often involves hormonal treatments to induce ovulation and the subsequent collection of mature eggs. Sperm is typically collected from males using non-invasive methods. The viability and quality of these gametes are critical for the success of IVF.
3. Egg and Sperm Preparation: Once collected, the eggs and sperm are prepared for fertilization. This preparation includes washing and concentration of sperm and the conditioning of eggs to enhance their fertilizability. This step is vital to ensure a high rate of successful fertilization.
4. Fertilization: The actual fertilization process involves mixing the sperm and eggs under controlled conditions. This is usually done in a petri dish or similar container, where the concentration of sperm and environmental conditions like temperature and pH are meticulously regulated to mimic natural fertilization as closely as possible.
5. Incubation: After fertilization, the embryos are incubated in a controlled environment. This stage is crucial for the development of the fertilized eggs. Conditions such as temperature, oxygen levels, and pH are closely monitored and adjusted to ensure optimal embryonic development.
6. Assessment of Fertilization: The fertilized eggs are assessed for signs of normal development. This assessment typically includes monitoring cell division and checking for any abnormalities in the early stages of embryonic growth.
7. Larvae Rearing: Once the embryos have developed into larvae, they are transferred to rearing tanks. In this stage, larvae are provided with appropriate nutrition and environmental conditions to ensure healthy growth and development.
8. Monitoring and Care: Throughout the larval stage, continuous monitoring and care are essential. This includes regular water quality checks, feeding, and health assessments to ensure the well-being and proper growth of the larvae.

12.4.1.3 Artificial Insemination

Artificial Insemination (AI) is a reproductive technique that involves the deliberate introduction of sperm into a female's reproductive tract for the purpose of achieving fertilization without natural mating. This process, crucial for controlled breeding and genetic management, entails several detailed steps:

1. Broodstock Selection: The process begins with the careful selection of male and female broodstock. These individuals are chosen based on specific criteria such as genetic quality, health, and desirable traits. This selection is fundamental to ensure that the offspring inherit the targeted characteristics.
2. Sperm Collection: Sperm is collected from the selected male fish. This usually involves manually stimulating the male to release sperm or extracting sperm from the male's reproductive organs. The collection process must be done gently to maintain the viability and quality of the sperm.
3. Sperm Analysis and Preparation: Once collected, the sperm undergoes analysis to assess its quality, including motility, concentration, and morphology. It is then prepared for insemination, which may involve dilution or the addition of extenders to preserve its viability.
4. Female Preparation: The female fish are prepared for insemination, often involving hormonal treatments to induce ovulation. Timing is critical; the eggs must be at the right stage of development for successful fertilization.
5. Insemination Process: The prepared sperm is then introduced into the female's reproductive tract. This can be done via various methods, depending on the species and the setup, such as direct injection into the ovary or near the genital opening.
6. Fertilization Monitoring: After insemination, the process of fertilization is monitored. This may involve observing the female for signs of successful fertilization or, in some cases, retrieving and examining the eggs to confirm fertilization.
7. Egg Incubation and Larval Rearing: Fertilized eggs are incubated under controlled conditions suitable for their development. Once hatched, the larvae are reared with appropriate care, including optimal feeding and maintaining water quality, to ensure their healthy growth.
8. Record Keeping and Analysis: Throughout the process, detailed records are kept for each step, including broodstock information, sperm quality, insemination details, and fertilization rates. This data is crucial for evaluating the success of the AI process and for making informed decisions in future breeding programs.

12.4.1.4 Cryopreservation

Cryopreservation in fisheries and aquaculture is a technique used to preserve gametes, embryos, or genetic material at ultra-low temperatures, typically involving liquid nitrogen. This process plays a crucial role in genetic conservation, selective breeding programs, and research. The detailed steps involved in cryopreservation are as follows:

1. Sample Collection and Preparation: The first step involves collecting the biological material to be cryopreserved, such as sperm, eggs, or embryos. The collected samples are then prepared, which may include cleaning, sorting, and treating them with cryoprotectants. Cryoprotectants are substances that protect the cells from damage during the freezing and thawing processes.
2. Cryoprotectant Treatment: After initial preparation, the samples are treated with cryoprotectants. These agents help in minimizing ice crystal formation within the cells, which can cause cellular damage. The choice and concentration of cryoprotectants are critical and vary depending on the type of cells being frozen.
3. Controlled Rate Freezing: The samples are then subjected to a controlled rate freezing process. This involves gradually lowering the temperature of the samples at a controlled rate to prevent thermal shock and minimize ice crystal formation. Controlled rate freezers are often used for this purpose.
4. Storage in Liquid Nitrogen: Once the samples reach a sufficiently low temperature, they are transferred to storage containers, typically dewars filled with liquid nitrogen, which maintains temperatures around −196 °C. At these temperatures, all biological activity, including cellular metabolism and genetic degradation, is effectively halted.
5. Record Keeping and Monitoring: Accurate record-keeping is essential for cryopreservation. It involves documenting the details of the samples, freezing protocols, cryoprotectant concentrations, and storage locations. Regular monitoring of storage conditions is also necessary to ensure the long-term viability of the samples.
6. Thawing and Reviving the Samples: When the samples are needed for use, they are rapidly thawed to minimize ice crystal formation during the warming process. This is typically done in a water bath at a controlled temperature. After thawing, the cryoprotectants are carefully removed from the samples, and the cells are gradually brought to physiological conditions.
7. Assessment of Viability: Post-thawing, the viability of the samples is assessed. This may involve checking the motility of sperm, the integrity of eggs or embryos, or conducting viability assays. The success of the cryopreservation process is largely dependent on the percentage of viable cells post-thaw.

Cryopreservation is a complex process requiring expertise in cryobiology and meticulous attention to detail. Its successful application in fisheries and aquaculture provides a valuable tool for genetic conservation, breeding program management, and research into aquatic species' biology and genetics.

12.4.1.5 Genetic Engineering and Gene Editing

Genetic engineering and gene editing involve the modification of an organism's genetic material to introduce desirable traits or remove unwanted characteristics. This advanced biotechnological process encompasses several steps:

1. Target Gene Identification: The initial step involves identifying and selecting the specific gene(s) to be edited or modified. This decision is based on the desired

outcome, such as improved growth rates, disease resistance, or adaptability to environmental conditions.

2. Vector Construction: Once the target gene is identified, a vector, typically a plasmid or a virus, is constructed. This vector will carry the desired genetic material or editing tools into the host organism's cells. In gene editing, this may involve the construction of a guide RNA (gRNA) that directs the gene editing tool to the specific location in the genome.
3. Gene Editing Tool Preparation: In gene editing, tools like CRISPR-Cas9 are prepared. CRISPR-Cas9, for instance, uses a gRNA to target a specific sequence in the genome and the Cas9 enzyme to cut the DNA at that precise location. This enables the addition, removal, or alteration of genetic sequences.
4. Transformation or Transfection: The next step is introducing the vector into the target organism's cells. This can be achieved through various methods like microinjection, electroporation, or viral transduction. The method chosen depends on the type of organism and cells being targeted.
5. Integration and Expression: Once inside the host cell, the genetic material carried by the vector integrates into the organism's genome. The inserted gene then begins to express, leading to the production of the desired protein or trait. In gene editing, the cell's natural repair mechanisms activate to repair the cut made by the gene editing tool, resulting in the modification of the genetic sequence.
6. Selection and Screening: Following transformation, cells or organisms that have successfully incorporated the genetic modification are selected and screened. This often involves using markers, like antibiotic resistance or fluorescent proteins, which were introduced along with the target gene.
7. Evaluation and Characterization: The genetically modified organisms are then evaluated to ensure the genetic modification is correctly expressed and functioning as intended. This evaluation includes molecular characterization, phenotypic analysis, and assessment of the trait of interest.
8. Breeding and Propagation: Once the desired genetic modification is confirmed and characterized, the modified organisms can be bred and propagated to produce offspring with the desired traits. This step is critical in aquaculture and fisheries for producing stocks with enhanced characteristics.

Genetic engineering and gene editing are highly technical and require sophisticated laboratory equipment and expertise in molecular biology and genetics. These technologies hold immense potential in improving the productivity, sustainability, and resilience of aquaculture and fisheries, contributing significantly to global food security and environmental conservation.

12.4.1.6 Hormonal Induction

Hormonal induction in fisheries and aquaculture is a process used to stimulate and control the reproductive cycle of fish and other aquatic species, particularly for breeding purposes. This technique involves administering hormones to induce spawning or the production of gametes (eggs and sperm). The detailed steps in hormonal induction are:

1. Broodstock Selection and Preparation: Initially, mature and healthy broodstock (male and female fish) are selected based on criteria such as age, size, and genetic qualities. These fish are often prepared for hormonal induction by ensuring optimal health and nutritional status, as these factors significantly influence the success of the induction process.
2. Determination of Reproductive Status: Before administering hormones, it's crucial to determine the reproductive status of the broodstock. This involves assessing factors like gonadal development, which can be done through physical examination, ultrasound imaging, or hormonal assays. Understanding the reproductive stage ensures the timely and effective application of hormonal treatments.
3. Hormone Selection and Dosage Calculation: The appropriate hormones and dosages are selected based on the species and the desired outcome. Common hormones used include gonadotropin-releasing hormone (GnRH), human chorionic gonadotropin (hCG), and luteinizing hormone (LH). The dosage is carefully calculated to optimize the spawning response while minimizing stress and potential side effects to the fish.
4. Administration of Hormones: The hormones are administered through injections or, in some cases, through water immersion. Injection methods are more precise and commonly used, with the hormone injected directly into the muscle or peritoneal cavity. The administration may occur in a single dose or multiple doses, depending on the species and hormone used.
5. Monitoring and Care Post-Induction: After hormonal induction, the fish are closely monitored for responses such as behavioural changes and signs of ovulation or spermiation. Environmental conditions like water temperature and quality are meticulously maintained, as they significantly affect the success of spawning.
6. Collection of Gametes: Once the fish exhibit signs of spawning readiness, gametes are collected. For females, this might involve stripping eggs, while males are often milked for sperm. This process needs to be gentle and efficient to ensure the quality and viability of the gametes.
7. Fertilization and Incubation: Following collection, eggs and sperm are fertilized, usually in a controlled environment. The fertilized eggs are then incubated under optimal conditions until they hatch into larvae.
8. Assessment and Evaluation: The success of hormonal induction is evaluated based on parameters like the quantity and quality of gametes produced, fertilization rates, and the health of the resulting larvae. This assessment is crucial for refining the hormonal induction protocols and improving future breeding efforts.

Hormonal induction is a critical tool in aquaculture and fisheries breeding programs, enabling the controlled reproduction of species that might not breed effectively in captivity. This process requires a deep understanding of endocrinology, reproductive biology, and species-specific requirements to be successfully implemented.

12.4.1.7 Embryo Transfer

Embryo transfer is a reproductive technology that involves transferring fertilized embryos from a donor fish to a recipient fish, which then carries the embryos to term. This method is particularly useful for breeding programs, genetic conservation, and research. The detailed steps involved in embryo transfer are as follows:

1. Donor and Recipient Selection: The process begins with the selection of suitable donor and recipient fish. Donors are chosen based on desirable genetic traits, while recipients are selected for their ability to successfully carry and hatch the embryos. The health and reproductive status of both donors and recipients are critical factors in this selection.
2. Superovulation of Donor: In many cases, the donor fish is treated with hormones to induce superovulation, which is the production of a larger than normal number of eggs. This increases the number of embryos available for transfer.
3. Collection of Eggs and Fertilization: Eggs are collected from the donor fish, often following hormonal induction of ovulation. These eggs are then artificially fertilized with sperm, either from a selected male in the breeding program or using cryopreserved sperm.
4. Incubation of Fertilized Eggs: Once fertilized, the eggs are incubated for a short period until they reach the appropriate stage for transfer. The incubation conditions, such as temperature and oxygen levels, are carefully controlled to ensure optimal embryo development.
5. Preparation of Recipient: The recipient fish is prepared to receive the embryos. This may involve treatments to synchronize its reproductive cycle with the developmental stage of the embryos, ensuring the recipient's body is ready to accept and nurture the transferred embryos.
6. Embryo Transfer Procedure: The embryos are carefully transferred into the reproductive tract of the recipient fish. This is a delicate procedure that requires precision to avoid damaging the embryos. The method of transfer varies depending on the species and the size of the embryos.
7. Post-Transfer Care and Monitoring: After the transfer, the recipient fish is monitored for signs of successful implantation and embryo development. This may involve regular health checks and, in some cases, non-invasive imaging techniques to observe the developing embryos.
8. Birth or Hatching: The recipient fish carries the embryos until they hatch. The hatching is monitored, and the health of the larvae is assessed. In some cases, the larvae may be genetically tested to confirm their parentage.
9. Record Keeping and Evaluation: Detailed records of the entire embryo transfer process are maintained, including information about the donor and recipient, the embryos transferred, and the outcomes. This data is crucial for evaluating the success of the procedure and for making improvements in future embryo transfer protocols.

Embryo transfer in aquaculture and fisheries is a complex technique that requires specialized skills and equipment. Its successful application can enhance breeding programs, aid in the conservation of endangered species, and contribute to research in aquatic biology and genetics.

12.4.2 Step-By-Step Process of Implementing these Technologies in Fisheries and Aquaculture

Proper implementation of these technologies related to surrogacy in fisheries and aquaculture is vital. In addition to this, the step-by-step planning and research is necessary to implement this in aquatic species. The steps in summary are as follows:

1. Objective Identification and Planning: The initial step involves identifying the specific objectives of implementing surrogacy technologies, whether it's for genetic improvement, species conservation, or research. Detailed planning is crucial, encompassing aspects such as species selection, resource allocation, and determining the appropriate surrogacy techniques to be used.
2. Facility and Resource Preparation: Setting up the appropriate facilities and ensuring the availability of necessary resources is vital. This includes creating suitable aquatic environments, acquiring the required technological and laboratory equipment, and ensuring a steady supply of materials like cryoprotectants, hormones, and genetic engineering tools.
3. Staff Training and Development: Given the technical complexity of surrogacy technologies, specialized training for staff is essential. This training covers aspects of reproductive biology, genetic engineering, embryo handling, and the specific techniques being implemented.
4. Broodstock Selection and Management: Selecting high-quality broodstock is fundamental in surrogacy programs. This involves choosing individuals with desirable genetic traits and ensuring their optimal health and nutrition, as this directly impacts the success of the reproductive processes.
5. Implementation of Specific Surrogacy Techniques: Depending on the objectives, specific surrogacy techniques are implemented. This may include hormonal induction for controlled breeding, germ cell transplantation, in vitro fertilization, artificial insemination, genetic engineering, or embryo transfer. Each technique requires meticulous execution following scientifically established protocols.
6. Monitoring and Data Collection: Continuous monitoring of the processes and outcomes is vital. This includes tracking the health and reproductive status of the broodstock, the condition of embryos or larvae, and the efficiency of the surrogacy techniques. Data collection is crucial for assessing the success of the program and for making informed adjustments.
7. Assessment and Quality Control: Regular assessment of the surrogacy processes and outcomes ensures adherence to quality and ethical standards. This includes evaluating the genetic integrity of the offspring, the health of the broodstock and surrogates, and the overall impact on biodiversity and the environment.

8. Research and Development: Ongoing research is integral to the advancement of surrogacy technologies. This involves staying updated with the latest scientific developments, experimenting with new techniques, and continuously improving existing methodologies.
9. Compliance and Ethical Considerations: Ensuring compliance with regulatory and ethical standards is essential. This involves adhering to guidelines for genetic manipulation, animal welfare, and environmental impact.
10. Outcome Analysis and Reporting: Finally, analysing the outcomes of the surrogacy program and reporting the results contribute to the broader understanding and advancement of these technologies. This includes publishing findings, sharing best practices, and participating in scientific discourse.

12.5 Ethical and Environmental Considerations

The integration of surrogacy technologies in fisheries and aquaculture raises several ethical and environmental considerations that necessitate careful deliberation. These concerns stem from the potential impacts of these technologies on animal welfare, genetic integrity, ecosystem balance, and broader ethical implications.

- Animal welfare: The use of surrogacy technologies often involves procedures like egg collection, hormonal treatments, and embryo manipulation, which must be conducted with utmost care to minimize stress and harm to the fish. Ethical considerations demand that these interventions adhere to the highest standards of animal welfare, ensuring that the procedures are as non-invasive and pain-free as possible.
- Genetic diversity and integrity: While genetic manipulation offers significant benefits, there's a concern about maintaining genetic diversity within species. Excessive focus on particular traits can lead to a reduction in genetic variability, potentially making populations more susceptible to diseases and environmental changes. Ensuring that genetic interventions do not compromise the genetic integrity and diversity of aquatic species is crucial.
- Impact on natural ecosystems: Releasing genetically modified or surrogate-bred fish into wild populations can have unforeseen consequences on natural ecosystems. There is a risk of these introduced species outcompeting native populations or altering the existing ecological balance. Rigorous risk assessments and controlled release strategies are essential to mitigate these environmental impacts.
- Ethical implications of genetic modification: The manipulation of the genetic makeup of organisms raises fundamental ethical questions. Concerns include the long-term implications of these modifications and the moral responsibility of altering living organisms at a genetic level. Ethical frameworks guiding these interventions need to be established, balancing scientific advancement with respect for the intrinsic value of natural life.
- Conservation vs. commercialization: While surrogacy technologies hold promise for species conservation, there's a tension between conservation goals and com-

mercial interests. Ensuring that these technologies are used primarily for sustainable and responsible aquaculture, rather than solely for profit-driven motives, is an ethical consideration that needs to be addressed.
- Public perception and acceptance: The acceptance of surrogate-bred or genetically modified aquatic species by consumers and the public is another crucial factor. Transparent communication regarding the safety, benefits, and environmental impacts of these technologies is necessary to build public trust and acceptance.
- Regulatory compliance: Adhering to legal and regulatory frameworks governing the use of biotechnologies in aquaculture is essential. These regulations are in place to ensure that the technologies are used safely, responsibly, and ethically, minimizing potential risks to both the environment and society.
- Sustainability: Finally, the sustainability of aquaculture practices using surrogacy technologies is a major consideration. This encompasses not just the environmental aspects, but also the social and economic dimensions, ensuring that these practices contribute positively to the communities involved and do not deplete natural resources.

12.6 Challenges and Limitations

Implementing surrogacy technology in fisheries and aquaculture presents several challenges and limitations, both technical and in terms of research and practice. These challenges often stem from the complex nature of biological systems, the intricacies of genetic manipulation, and the practical aspects of working within aquatic environments. These challenges can be grouped into two categories for better understanding, i.e., technical challenges and limitations in current research and practice.

12.6.1 Technical Challenges in Implementing Surrogacy Technology

1. Complexity of genetic manipulation: Genetic engineering and editing require precise manipulation of the genome, which is a highly complex and delicate task. Achieving targeted and error-free genetic alterations without off-target effects or unintended consequences remains a significant technical challenge.
2. Broodstock and recipient compatibility: In practices like germ cell transplantation or embryo transfer, finding compatible broodstock and recipient species can be difficult. The physiological and immunological compatibility between donor and recipient is crucial for the success of these procedures.
3. Controlled reproduction conditions: Creating optimal conditions for reproduction, including hormonal induction, artificial insemination, or IVF, requires meticulous control of environmental factors. Achieving the right balance of conditions to mimic natural breeding can be technically challenging.

4. Cryopreservation and thawing: While cryopreservation is a powerful tool for conserving genetic material, the freezing and thawing processes can be technically demanding. Ensuring the viability of gametes or embryos post-thaw requires precise control of temperatures and cryoprotectant concentrations.

12.6.2 Limitations in Current Research and Practice

1. Limited understanding of reproductive biology: Despite advances, there is still a limited understanding of the reproductive biology of many aquatic species. This gap in knowledge can hinder the effective application of surrogacy technologies, especially for less-studied or wild species.
2. Ethical and regulatory constraints: Ethical concerns and regulatory restrictions regarding genetic manipulation can limit the scope of research and application of surrogacy technologies. Navigating these ethical and regulatory landscapes can be challenging, especially in regions with strict controls on genetic engineering.
3. Risk of genetic homogenization: There is a risk that extensive use of surrogacy technologies could lead to genetic homogenization, reducing the genetic diversity within species. This limitation is particularly relevant in selective breeding programs focused on specific traits.
4. Environmental impact assessment: Assessing the long-term environmental impacts of introducing surrogate-bred or genetically modified organisms into ecosystems is a significant research challenge. Understanding these impacts is crucial for the responsible application of these technologies.
5. Scalability and economic viability: Scaling up surrogacy technologies for commercial aquaculture can be challenging. The costs and complexities of implementing these technologies at a larger scale may not always be economically viable, particularly for smaller operations.
6. Public perception and market acceptance: Public perception and market acceptance of products derived from genetically manipulated organisms can be unpredictable. Negative public attitudes or market resistance can limit the practical application of these technologies.

12.7 Future Perspectives

The future of surrogacy technology in fisheries and aquaculture is poised for significant advancements, shaped by emerging trends and an expanding scope for innovation. As we delve into the future perspectives, it's clear that this field stands at the cusp of transformative changes, driven by scientific breakthroughs and the growing necessity for sustainable aquaculture practices. One of the prominent emerging trends is the increasing sophistication of genetic engineering and gene editing techniques. Tools like CRISPR-Cas9 are revolutionizing the way genetic modifications are made in aquatic organisms. Future directions point towards more precise, efficient, and safe methods of gene editing, potentially enabling the creation of fish breeds with enhanced growth rates, superior disease resistance, and better adaptability to changing environmental conditions. These advancements promise not only to boost aquaculture productivity but also to contribute to more sustainable and environmentally friendly practices.

Another trend is the integration of artificial intelligence (AI) and machine learning in surrogacy technologies. These technologies can significantly enhance the efficiency of breeding programs, from predicting optimal breeding pairs to monitoring the health and growth of surrogate-bred organisms. AI-driven data analysis can provide deeper insights into genetic traits and their expression, paving the way for more informed decision-making in genetic manipulation and breeding strategies. Advances in cryopreservation techniques hold substantial potential for the future. Innovations in this area could lead to more effective and reliable methods of preserving genetic material, ensuring the long-term viability of gametes and embryos. This has profound implications for biodiversity conservation, allowing for the preservation of endangered species and the maintenance of genetic diversity within cultured populations. Aquaculture systems themselves are likely to undergo significant evolution. The integration of surrogacy technologies with closed-loop and recirculating aquaculture systems (RAS) can lead to highly efficient, controlled, and sustainable aquaculture practices. These systems, combined with advanced reproductive technologies, could revolutionize the way fish are bred and grown, minimizing environmental impacts and maximizing production efficiency. There is also growing interest in developing surrogacy technologies for a wider range of species, including those that are currently less represented in aquaculture. This expansion could open new markets and opportunities, contributing to the diversification of the aquaculture industry and providing new sources of nutritious and sustainable seafood. Additionally, the field of regenerative aquaculture, where surrogacy technologies are used in conjunction with habitat restoration efforts, presents a promising direction. This approach not only focuses on breeding and production but also on the ecological restoration of aquatic environments, contributing to the overall health of marine ecosystems.

12.8 Conclusion

In conclusion, the realm of surrogacy technology in fisheries and aquaculture represents a remarkable amalgamation of biotechnology, genetics, and ecological science. It stands as a testament to the ingenuity and resourcefulness of scientific endeavours aimed at addressing the complex challenges faced by the aquaculture industry and the conservation of aquatic biodiversity. Through the application of advanced techniques such as genetic engineering, germ cell transplantation, and hormonal induction, significant strides have been made in enhancing fish breeding, improving species resilience, and ensuring sustainable food sources. However, this technological landscape is not without its challenges and ethical considerations. The meticulous balance between advancing aquacultural productivity and maintaining ecological integrity remains a critical focus. As we navigate the intricacies of genetic manipulation and the ecological impacts of these technologies, a continuous and informed dialogue is essential, encompassing scientific, ethical, and environmental perspectives. Looking forward, the potential for innovation in surrogacy technology is immense. Emerging trends, particularly in genetic editing and artificial intelligence, are set to redefine the boundaries of what can be achieved in fisheries and aquaculture. These advancements, coupled with a growing emphasis on sustainability and ecological responsibility, herald a new era in the field. The convergence of scientific progress with sustainable practices offers a hopeful vision for the future, promising to address global food security challenges, conserve endangered species, and foster a harmonious relationship with our aquatic ecosystems.

References

Almendras JM, Duenas C, Nacario J, Sherwood NM, Crim LW (1988) Sustained hormone release. III. Use of gonadotropin releasing hormone analogues to induce multiple spawnings in sea bass, Lates calcarifer. Aquaculture 74:97–111

Baloch AR, Franěk R, Tichopád T, Fučíková M, Rodina M, Pšenička M (2019) Dnd1 knockout in sturgeons by CRISPR/Cas9 generates germ cell free host for surrogate production. Animals 9:174

Biswas JK, Maurye P (2017) Aquaculture biotechnology: prospects and challenges. Progress Biotechnol India 1:33–47

Chavanne H, Janssen K, Hofherr J, Contini F, Haffray P, Consortium A, Komen H, Nielsen EE, Bargelloni L (2016) A comprehensive survey on selective breeding programs and seed market in the European aquaculture fish industry. Aquac Int 24:1287–1307

Cowl VB, Comizzoli P, Appeltant R, Bolton RL, Browne R, Holt WV, Penfold LM, Swegen A, Walker SL, Williams SA (2023) Cloning for the twenty-first century and its place in endangered species conservation. Ann Rev Anim Biosci 12:91

Dehler CE, Boudinot P, Martin SA, Collet B (2016) Development of an efficient genome editing method by CRISPR/Cas9 in a fish cell line. Mar Biotechnol 18:449–452

Donaldson LR (1971) Selective breeding of salmonid fishes, proceedings of the annual workshop-world mariculture society, vol 2. Wiley, New York, pp 75–83

Donaldson LR, Olson PR (1957) Development of rainbow trout brood stock by selective breeding. Trans Am Fish Soc 85:93–101

Dunham RA (2023) Aquaculture and fisheries biotechnology: genetic approaches. CABI, Boston

Franěk R, Cheng Y, Fučíková M, Kašpar V, Xie X, Shah MA, Linhart O, Šauman I, Pšenička M (2022) Who is the best surrogate for germ stem cell transplantation in fish? Aquaculture 549:737759

Gambini A, Ortiz I, Hidalgo M, Dorado J (2023) Selection and conservation of gametes and embryos for improving assisted reproductive technologies. Animals 13:2761

Garber AF, Sullivan CV (2006) Selective breeding for the hybrid striped bass (Morone chrysops, Rafinesque× M. Saxatilis, Walbaum) industry: status and perspectives. Aquac Res 37:319–338

Gazsi G, Ivánovics B, Izabella RB, Szabó T, Daniel Z, Kucska B, Urbányi B, Horváth L, Müller F, Müller T (2021) Artificial sperm insemination in externally fertilised fish as a novel tool for ex situ and in situ conservation of valuable populations. Endanger Species Res 45:169–179

Gjedrem T, Robinson N, Rye M (2012) The importance of selective breeding in aquaculture to meet future demands for animal protein: a review. Aquaculture 350:117–129

Gui J-F, Zhou L, Li X-Y (2022) Rethinking fish biology and biotechnologies in the challenge era for burgeoning genome resources and strengthening food security. Water Biol Secur 1:100002

Jin YH, Robledo D, Hickey JM, McGrew MJ, Houston RD (2021) Surrogate broodstock to enhance biotechnology research and applications in aquaculture. Biotechnol Adv 49:107756

Khosla K, Wang Y, Hagedorn M, Qin Z, Bischof J (2017) Gold nanorod induced warming of embryos from the cryogenic state enhances viability. ACS Nano 11:7869–7878

Lacerda S, Costa G, Campos-Junior P, Segatelli T, Yazawa R, Takeuchi Y, Morita T, Yoshizaki G, França L (2013) Germ cell transplantation as a potential biotechnological approach to fish reproduction. Fish Physiol Biochem 39:3–11

Leberg PL (1992) Effects of population bottlenecks on genetic diversity as measured by allozyme electrophoresis. Evolution 46:477–494

Liu Q, Yuan Y, Zhu F, Hong Y, Ge R (2018) Efficient genome editing using CRISPR/Cas9 ribonucleoprotein approach in cultured Medaka fish cells. Biol Open 7:bio035170

Lujić J, Marinović Z, Bajec SS, Djurdjevič I, Urbányi B, Horváth Á (2018) Interspecific germ cell transplantation: a new light in the conservation of valuable Balkan trout genetic resources? Fish Physiol Biochem 44:1487–1498

Majhi SK, Hattori RS, Yokota M, Watanabe S, Strüssmann CA (2009) Germ cell transplantation using sexually competent fish: an approach for rapid propagation of endangered and valuable germlines. PLoS One 4:e6132

Mañanós E, Duncan N, Mylonas C (2008) Reproduction and control of ovulation, spermiation and spawning in cultured fish. In: Methods in reproductive aquaculture: marine and freshwater species. CRC Press, Boca Raton, pp 3–80

McCusker MR, Bentzen P (2010) Positive relationships between genetic diversity and abundance in fishes. Mol Ecol 19:4852–4862

Mitra A, Abdel-Gawad FK, Bassem S, Barua P, Assisi L, Parisi C, Temraz TA, Vangone R, Kajbaf K, Kumar V (2023) Climate change and reproductive biocomplexity in fishes: innovative management approaches towards sustainability of fisheries and aquaculture. Water 15:725

Mobrand LE, Barr J, Blankenship L, Campton DE, Evelyn TT, Flagg TA, Mahnken CV, Seeb LW, Seidel PR, Smoker WW (2005) Hatchery reform in Washington state: principles and emerging issues. Fisheries 30:11–23

Muñoz-Cueto JA, Zmora N, Paullada-Salmerón JA, Marvel M, Mañanos E, Zohar Y (2020) The gonadotropin-releasing hormones: lessons from fish. Gen Comp Endocrinol 291:113422

Mylonas CC, Fostier A, Zanuy S (2010) Broodstock management and hormonal manipulations of fish reproduction. Gen Comp Endocrinol 165:516–534

Segev-Hadar A, Slosman T, Rozen A, Sherman A, Cnaani A, Biran J (2021) Genome editing using the CRISPR-Cas9 system to generate a solid-red germline of Nile tilapia (Oreochromis niloticus). CRISPR J 4:583–594

Yoshizaki G, Lee S (2018) Production of live fish derived from frozen germ cells via germ cell transplantation. Stem Cell Res 29:103–110

Yoshizaki G, Yazawa R (2019) Application of surrogate broodstock technology in aquaculture. Fish Sci 85:429–437

The manufacturer's authorised representative in the EU is Springer Nature Customer Service Centre GmbH, Europaplatz 3, 69115 Heidelberg, Germany. If you have any concerns regarding our products, please contact ProductSafety@springernature.com

Printed and bound by CPI Group (UK) Ltd, Croydon, CR0 4YY
01/10/2026
02225863-0003